CONCEPTIONS IN THE CODE

OXFORD STUDIES IN LANGUAGE AND LAW

Oxford Studies in Language and Law includes scholarly analyses and descriptions of language evidence in civil and criminal law cases as well as language issues arising in the area of statutes, statutory interpretation, courtroom discourse, jury instructions, and historical changes in legal language.

The Legal Language of Scottish Burghs: Standardization and Lexical Bundles (1380–1560)
Joanna Kopaczyk

"I'm Sorry for What I've Done": The Language of Courtroom Apologies
M. Catherine Gruber

Dueling Discourses: The Construction of Reality in Closing Arguments
Laura Felton Rosulek

Entextualizing Domestic Violence: Language Ideology and Violence Against Women in the Anglo-American Hearsay Principle
Jennifer Andrus

Speak English or What? Codeswitching and Interpreter Use in New York City Small Claims Court
Philipp Sebastian Angermeyer

Law at Work: Studies in Legal Ethnomethods
Edited by Baudouin Dupret, Michael Lynch, and Tim Berard

Speaking of Language and Law: Conversations on the Work of Peter Tiersma
Edited by Lawrence M. Solan, Janet Ainsworth, and Roger W. Shuy

Confronting the Death Penalty: How Language Influences Jurors in Capital Cases
Robin Conley

Discursive Constructions of Consent in the Legal Process
Edited by Susan Ehrlich, Diana Eades and Janet Ainsworth

From Truth to Technique at Trial: A Discursive History of Advocacy Advice Texts
Philip Gaines

Discourse, Identity, and Social Change in the Marriage Equality Debates
Karen Tracy

Translating the Social World for Law: Linguistic Tools for a New Legal Realism
Edited by Elizabeth Mertz, William K. Ford, and Gregory Matoesian

CONCEPTIONS IN THE CODE

How Metaphors Explain Legal Challenges in Digital Times

Stefan Larsson

OXFORD
UNIVERSITY PRESS

OXFORD
UNIVERSITY PRESS

Oxford University Press is a department of the University of Oxford. It furthers the University's objective of excellence in research, scholarship, and education by publishing worldwide. Oxford is a registered trade mark of Oxford University Press in the UK and certain other countries.

Published in the United States of America by Oxford University Press
198 Madison Avenue, New York, NY 10016, United States of America.

Library of Congress Cataloging-in-Publication Data
Names: Larsson, Stefan, 1978– author.
Title: Conceptions in the code : how metaphors explain legal challenges in digital times / Stefan Larsson.
Description: New York, NY : Oxford University Press, 2017. |
Includes bibliographical references and index.
Identifiers: LCCN 2016021520 (print) | LCCN 2016021993 (ebook) |
ISBN 9780190650384 (hardback : alk. paper) | ISBN 9780190650391 (pdf) |
ISBN 9780190650407 (e-book)
Subjects: LCSH: Intellectual property—Social aspects. | Peer-to-peer architecture (Computer networks)—Law and legislation—Social aspects. |
Technological innovations—Law and legislation—Social aspects.
Classification: LCC K1401 .L37 2017 (print) | LCC K1401 (ebook) |
DDC 346.04/8—dc23
LC record available at https://lccn.loc.gov/2016021520

1 3 5 7 9 8 6 4 2

Printed by Sheridan Books, Inc., United States of America

CONTENTS

ACKNOWLEDGMENTS

This book is in many senses of the word a remix. First of all, it is so because several of the chapters originate from previously published journal articles that I have revised to fit the grander framework of a book. Secondly, however, it is also a remix in the somewhat less acknowledged sense that the contributions to any creative output by necessity comes from a multitude of sources, and depend on context and culture. Just because my name is on the cover by no means mean that I was alone to produce the text or the ideas it describes. In this case, firstly, I wish to express my gratitude to all the (mostly) anonymous peers that have reviewed the preceding articles as well as journal editors. They have all contributed to increase the quality of my statements and the coherence of my arguments. This gratitude is of course extended to the reviewers of an earlier version of this book, which significantly facilitated necessary clarifications and the progression of the book, as well as editor Hallie Stebbins at Oxford University Press.

The dependence of authorship on forerunners and context is in a sense acknowledged in the conception of science expressed in the metaphor of 'standing on the shoulders of giants', but it is not so much acknowledged in the sense that authorship tends to be regarded as an expression of a single inspired author (hence the strong copyrights). Much of this book truly expresses a reliance on already present or past giants, and I have had inspired input from great minds amongst the sociolegal classics as well as the scholarly communities of cognitive theory and law, and some of which I have had the chance to meet, if only briefly, such as George Lakoff and Lawrence Lessig. Furthermore, I wish to thank my colleagues at Lund University, working with or close to the Internet Institute, perhaps especially Måns Svensson, a colleague and friend who always stands for a fearless and pioneering attitude, in life and in science. A special thank you also goes out to all the contributors to the studies conducted for this book, which includes file-sharers in all corners of the world responding to surveys as well as The Pirate Bay itself, whatever you are. Furthermore, this book was written and assembled during a time where I was a visiting researcher at a number of places—I've been blessed, really—for which I am very grateful to have welcomed me, such as the Department of Political Science at the University of Hawaii at Mānoa (thank you, Debora Halbert et al.); The IT-University of Copenhagen (thank you, Gitte Stald et al.); the Centre for Comparative and European Constitutional Studies (CECS) at the Faculty of Law, University of Copenhagen (thank you, Helle Krunke and Hanne Petersen et al.); and the Alexander von Humboldt Institute for Internet and Society (HIIG) in Berlin (thank you, Karina Preiss et al.). This also includes the Pufendorf Institute for Advanced Studies at Lund University, which funded and hosted our interdisciplinary research theme on digital trust and the digital society, in which we debated many metaphors (thank you, Per Runeson et al.); as well as the Swedish Research Council, making this research possible in the first place, through its beneficial funding. And finally, to Anna and the boys, thank you for all the support and bringing of balance to my life.

CONCEPTIONS IN THE CODE

[1]

HOW WE UNDERSTAND TECHNOLOGICAL AND SOCIAL CHANGE

Poets, deserted by the world before,
Turn round into the actual air:
Invent the age! Invent the metaphor![1]

In 2009, during the Swedish trial of the founders of the The Pirate Bay (TPB)—sometimes called the most resilient BitTorrent file-sharing site in the world—there was a sequence when the prosecutor asked the defendant Peter Sunde about the first time the defendants met IRL, an Internet slang for 'in real life'. Sunde replied that they do not use that terminology, 'We use AFK, meaning "away from keyboard", because we think that the Internet is for real.'

In this exchange, we see an indication of two completely different conceptions of the Internet. The prosecutor sees the Internet as something apart from and inferior to the 'real' world. The defendant, in contrast, experiences the Internet as an ordinary part of the social world that just happens to be accessed

1. The poem is part of Archibald MacLeish's 'Hypocrite Auteur' from *Collected Poems 1917–1982* (Boston: Houghton Mifflin, 1985).

Conceptions in the Code. Stefan Larsson.

through a keyboard rather than via a phone, a letter, or face-to-face.[2] That there are differing ways of understanding the Internet, and that they have implications for law and regulation, have been analysed in previous research. For example, the American legal scholars Jonathan Blavin and Glenn Cohen (2002) have studied three different conceptualisations of the Internet and what they entail for law: the *information superhighway*, the *cyberspace* as a distinct other space, and the *Internet as real space*. They found that conceptual metaphors for the Internet significantly impact how it is framed in legal contexts.

In this book I argue, too, that the conceptual battles to define the Internet, as well as the implications of digital development, are significant battles for the role of law in society. Metaphorical descriptions of the Internet and the significance of digitalisation are not just rhetoric or poetic language, but an outcome of how we understand this complex digital phenomenon and how this understanding can be negotiated, over time and across cultures, with strong discursive relevance. In brief, how we regulate the Internet, respond to its regulation, and behave online, is arguably linked to how we metaphorically understand the digital domain with all its inherent phenomena and functions.

Without concepts and imagery, we cannot speak about or understand the new, and without metaphors we have no concepts for new phenomena. Just consider the following metaphors: the Cloud, networks, streaming, data-flow, desktop, interface. And note, at the same time, how strongly linked these concepts are to a domain that is not digital or part of the Internet at all. The point here is that these labels transfer something from the

2. The documentary filmmaker Simon Klose also chose the latter terminology for the title of his film *TPB AFK* (2013), largely focusing the court case and the three most central defendants in the case.

source domain to the target domain of the new phenomena they describe, which controls our understanding of them, as well as our regulation of them.

This book uses conceptual metaphor theory, that is, a specific strand within cognitive theory, to strengthen the awareness of the strong metaphoricity in contemporary understandings of the Internet. The particular case used here concerns copyright and the challenges it meets in a digital context—often denoted 'file-sharing' or 'piracy'—and especially debated during the first decade or so of the new millennium. The studies conducted are both empirical surveys of the file-sharing community as well as studies of the use of certain, legally relevant terminology in published texts over time. This is complemented by conceptual analyses of both the Swedish court case concerning the founders of TPB as well as specific key terminology in copyright law, such as 'copy' and 'property'. The need to improve legal awareness of how metaphors function, particularly with regards to how new technology is conceptualised, is concisely put by Blavin and Cohen: 'By failing to adopt appropriate metaphors in regulating new technologies, courts risk creating bad law' (2002, p. 268). The purpose of this book is somewhat broader, and takes a step further from the work of courts to include how concepts in legislation are reinterpreted in relation to new technological phenomena—that is, to shed light on abstract thinking with regards to both law and technology, and how they relate in times of social change. Nevertheless, to avoid 'bad law' is of course still a great motivation for this. In brief, the heart of the issue addressed in this book is the role that metaphor plays when the law is dealing with technological change.

The epithet we give to an age tells us something not only about what we feel should be particularly highlighted or particularly avoided, but influences our thoughts on the subject as well. The images—metaphors—are powerful cognitive phenomena that we depend on to describe phenomena as abstract, yet malleable, as an age. In descriptions of the Internet as an *information superhighway* (cf. McCoy and Needham, 1995; Blavin and Cohen, 2002), the invoked imagery leads to a conceptual understanding that lies close to notions of transferring information from A to B—a concise, simple metaphor, that according to Blavin and Cohen (2002) leads to a suitability for state involvement.[3] When William Gibson coined the term *cyberspace* in his novel *Neuromancer* (1984), he conveyed the imagery that the early PCs and arcade games shared a mutual space that stretched beyond, or behind, the screens (Greiffenstern, 2010, p. 123; cf. Blavin and Cohen, 2002, pp. 275–80). This concept emphasised the idea that digitalised networks are spatial and have three dimensions that one can exist and move within, as well as an environment distinctly different from 'the real' world, and one may remember the early days of 'virtual reality' in the 1990s or the media attention in 2005–2006 for the 'virtual world' *Second Life*. This essence of difference was also an imperative part of early visions of the Internet and its relationship to regulation—a 'space' that should not, and even could not, be targeted by regulatory efforts under government. This is perhaps most famously expressed in John Perry Barlow's *A Declaration of the Independence of Cyberspace*, a text that was published online in 1996 as a response to the passing into law of the Telecommunications Act of 1996 in the United States. Barlow declared, addressing the governments of the industrial world as 'weary giants of flesh and steel', that 'cyberspace does not lie within your borders' and 'we are forming our own social

3. European Commission, Green Paper of 27 July 1995 on Copyright and Related Rights in the Information Society—COM(95)/0382 final.

contract. This governance will arise according to the conditions of our world, not yours. Our world is different.'

The notion of cyberspace as a differently regulated space found support in scholars like David R. Johnson and David Post, in articles like *Law and Borders—The Rise of Law in Cyberspace* from 1996. The unregulated aspects of 'cyberspace' did however not stick for very long, but the notion of Internet-related phenomena as a 'space' remained, as something qualitatively different from the traditional and analogue world, often argued to be rendering a particular type of regulatory means. The specifics of cyber-governance became a hot topic in the late 1990s. The Internet as a regulated (and real) space was argued for not the least by Lawrence Lessig, who formulated a widely known critique of the cyber-anarchistic approach expressed by Barlow (Lessig 1999, 2006). Lessig described the 'regulability' of the Internet and developed the innovative perspective of regulating *by* code, and described the Internet in terms of layers 'that are the Internet' (Lessig, 2006) such as the data link layer, the network layer, the transport layer, and the application layer (and these operate above 'a physical layer'). This imagery is supported by, for example, the writer and documentary filmmaker Astra Taylor. In *The People's Platform* she argues that 'the Internet is best thought of as a series of layers: a physical layer, a code layer, and a content layer' (2014, p. 36). Examples of understanding the coded architecture as something that is by necessity regulated—and in fact *a regulator*—is seen not the least in the calls for privacy *by design*, as opposed to merely protecting privacy *by policy* (cf. Halbert and Larsson, 2015; Larsson, 2013e). In this perspective the Internet is something very real, and the layered imagery acknowledges the physicality of the Internet as a kind of foundation for the communicative applications built 'on top' of it—with evident links to how regulatory debates around digital phenomena are shaped, as well as how social norms concerning digitally mediated behaviour develop.

The software code in itself, so to speak, is part in a grand assemblage of how computers work, that in the shape of software is particularly hard to pinpoint ontologically. Software is, as expressed by van den Boomen, 'in principle soft and fluid, ever open to adding, modifying, and extending' (2014, p. 128), and a key argument here is that the design it is part of also can be laden with values, making it important to be aware of precisely which values it enables or lead to.

> We can build, or architect, or code cyberspace to protect values that we believe are fundamental. Or we can build, or architect, or code cyberspace to allow those values to disappear. There is no middle ground. (Lessig, 2006, p. 6)

The argument in this book goes somewhat further in its cognitivist approach: how we understand phenomena will also be coded, and how the conceptualisations of specific phenomena will be part in the building and architecture of 'cyberspace'. Or, to contextualise it more, the conceptions of those who construct the digitised realm, including the code, will guide this construction. Many of the underlying values will arguably be depending on the conceptions, which therefore need to be unlocked.

This book's point of departure, then, lies in asking how, and why, if one wishes to understand legal developments in relation to new technology, it is of such fundamental importance to understand metaphors and conceptual change. One should bear in mind that technology never simply *is*—it is always interpreted within the context and society that it arises and exists in, and thereby becomes dependent on older technologies and already existing concepts that we adopt to understand the new. At the same time, new technological innovations—and many other digital phenomena—can radically alter the conditions of things that are important to us: how we communicate, how we participate in society, how we learn, how we consume, and how markets function. These changes can occur rapidly, as digitalisation has done, and permutations can flow over us quickly, as the spread of smartphones, wearables and tablet computers has done—in other words, placing the Internet in our hands and on our bodies. Law is however generally not understood as a swiftly changing phenomenon that quickly adapts to a changing context. And although there are many advantages to legal developments lagging somewhat—predictability is an important component of law and order—there are also disadvantages to a legal development that risks being too retrospective, conservative, and path dependent. Particularly, perhaps, in the case of technology, friction may arise when the regulating concepts have been framed in an era that has long passed, shaped by conditions that once prevailed but do so no longer. Much of this can arguably be attributed to how we understand new phenomena and the new conditions they produce, and to the concepts we use to recognise and communicate them.

Metaphors are relevant in the context of technology and law, not the least due to them demonstrably being very important for how we understand events such as legal constructions as well as digital phenomena. It is important to remember—and here we touch upon the essence of this book—that metaphors, although they are often a method of consciously rephrasing a phenomenon to achieve a certain effect (for instance through colourful images) are equally often an articulation of a process that is somewhat obscured from our consciousness; a mental model that quietly hinders us from thinking in certain modes as much because it encourages us to think in other given modes. In this manner, these mental models guide our thoughts without us really being aware of them. By pointing out *how* we understand abstract phenomena metaphorically, particularly with regards to how we express patterns or recurring structures, we may unravel some of the legal challenges that we have to tangle with as a result of rapid technology-related change. This, then, is this book's promise.

CONCEPTUAL STRUGGLES IN SOCIETAL CHANGE

The analysis of conceptual metaphors is relevant for a broader phenomenon of how society is appropriating or adapting to new technology. The concepts are neither neutral, objective, nor superficial. Copyright is here used as a case to prove this point, and it draws from multiple studies conducted in the first decade or so of the millennium where online file-sharing was intensely debated from a legal, entrepreneurial, and social point of view.

Contemporary conceptual struggles around big data and the data practises of massive collection, analysis, and brokerage show similar traits. Depending on how the issue is framed—the data practises utilise 'a resource' or drive the 'surveillance' of individuals—it can be a sign of openness or a threat to proprietary structures. Much of the new services, the way that consumers are profiled or security services operate, revolve around massive human-centric information gathering, i.e. the data that arguably is 'Big'. Sometimes described as the 'oil' of the new economy, as a resource to be harnessed, and sometimes described as a fluid entity, uncontrolled and overwhelming (cf. Puschmann and Burgess, 2014). I will briefly touch upon this development and its relation to the copyright discourse in the last chapter of the book.

New metaphors guide our understanding of digitally related phenomena. Some of these become commonplace, but still affect or shape how we conceptualise a number of key activities of modern society. For example, it is hard to talk about contemporary security issues without having to deal with concepts such as 'cloud', 'virus', 'leak', or 'hack'. It is hard to talk about contemporary entrepreneurship and innovation without 'disruption', 'sharing economy', or 'openness'; it is hard to talk about the role of media in a digital context without concepts such as 'paywall', 'streams', or even 'cookies'. And, it is most likely impossible to talk about file-sharing without the metaphors that organise it, such as 'torrents', 'swarms', 'trackers', or 'magnet links'. Similarly, you can't analyse copyright's modern challenges without talking about file-sharing, or deconstructing what concepts like 'stealing', 'trespassing', and 'piracy' actually bring to the debate.

In short, digital development means a whole new lexicon of concepts that all relate to an already present conceptual domain. A reflection here is that metaphor struggles may represent important points in time for societal change, where aspects of technology, innovation, business, economy, politics, law, and the social realm grind their understandings around particularly challenging events.

NEUTRAL INFRASTRUCTURE OR FILTERING MEDIATOR?

Some metaphors are explicit and clearly laden with values, such as Turkey's Prime Minister Tayyip Erdoğan's statement at an election meeting in March 2014: 'We will rip out Twitter by the roots and erase it!', following demonstrations that had been coordinated via social media and the spread of audio recordings damaging the government's reputation. That there are values expressed here are relatively blatant. But such metaphors can also act implicitly and not wholly consciously, and express a thought pattern rather than a rhetorical effect. Nevertheless, they may still remain value-based or lead the mind in a certain normative direction. It is at this point, when we are led without realising it, that it becomes even more crucial to place a spotlight on how these processes occur; which underlying images, conceptual mappings, and thought structures are leading us; and how these have come to emerge in the first place.

Returning to the topic of technology, much of the struggles that follow from the challenges posed by digitalisation concern responsibility and accountability, i.e. who is seen as responsible and what they should be held accountable for. And this responsibility is often linked to whether the service being used is perceived as passive or active. Some concepts lead us in one direction, while others lead in another. 'Infrastructure', for example, is a concept commonly used to indicate services—a 'mere conduit'—where, it is argued, the service providers cannot be held wholly responsible for events that occur there—much as if they merely provided a railway for trains to run on, or a postman that delivers the post. Internet service providers have stood in the line of fire for many years, in a struggle over whether or not they should be held accountable for copyright infringements caused by their users' online actions.[4] Another issue that has been long debated

4. How the service provider handled their customers' data was also the topic of much debate when the Data Retention Directive was rejected by the European Court of Justice (Ledendal and Larsson, 2014).

concerns the degree to which the service providers should be allowed to prioritise data in Internet traffic, which is a highly advantageous option for competitive services such as streaming media. To conceptually portray Internet service providers as *an infrastructure*, however, indicates that no priorities should be permissible—all traffic should be treated equally, and no traffic should be controlled. The subscribers should be responsible for their own actions. As mentioned previously, Turkey's Prime Minister at the time was seemingly of the opinion that the actual Twitter service itself bears responsibility for the criticism expressed by its users. Apparently he does not conceptualise it as a *passive* infrastructure.

'Platform' is another interesting concept that commonly crops up in digital contexts. It represents what communications and information science researcher Tarleton Gillespie refers to as a structural metaphor which 'though it may go unnoticed by the casual listener or even the speaker, gives the term discursive resonance' (Gillespie, 2010). The above mentioned Astra Taylor states that the concept 'implies a smooth surface, misleads us, obscuring the ways technology companies shape our online lives, prioritising and upraising certain purposes over other' (2014, p. 139). She has new media giants like Facebook and Google, to which she devotes much attention, in mind. This is also echoed in the recent increase in talk of a 'platform economy', sometimes shaped as a critique against the notion of a 'sharing economy' (Kenney and Zysman, 2016). There is, then, a sort of doublesidedness to the concept; The 'platform' comes with a notion of neutrality and even passivity, but the arguments against this notion in its digital version often claim that it still means some sort of activity in that it may monitor, sort or moderate the activities catalysed by the 'platform'.

One legal case that serves as an example of this argumentative battle around a passive/active duality concerns the court case against the founders of the file-sharing service TPB (further developed in chapter 6). Here, the key issue of establishing the TPB founders' degree of accountability for the illegal file-sharing facilitated by the site was linked to defining TPB itself—a task with highly conceptual and metaphorical traits. Was TPB a platform, a storage service or a bulletin board? To what extent was it a search engine? To what extent were the founders entrepreneurial? However, on 17 April 2009, four men were sentenced to one year's incarceration as well as being found jointly and severally responsible for damages of roughly EUR 3,500,000 for complicity in copyright infringements. Both parties appealed and the three defendants present during proceedings in the Court of Appeal were sentenced in the autumn of 2010 to shorter prison sentences, but increased damages of roughly EUR 5,400,000.[5] This specific case contains a number of very important questions concerning the point of intersection between social change, technical developments, and cognitive understanding of digital phenomena, and actual law itself (cf. Larsson, 2013c).

During the case, defence lawyer Jonas Nilsson stated that 'The only thing The Pirate Bay does is passively provide an empty platform'.[6] One of the American plaintiff's lawyers, Monique Wadsted, stated during the Court of Appeal proceedings that TPB was not a 'passive web site' and argued that the hyperlinks to different categories on the web site were an instance of simplifying the web site interface for the benefit of the users, which would

5. The Court of Appeal settled the district court's verdict for the fourth defendant, who was taken ill abroad at the time.

6. During day twelve at the District Court. Sverige Radio. 3 March, 2009. http://sverigesradio.se/sida/artikel.aspx?programid=83&artikel=2672385

contradict claims of 'passivity' (See SvD's report, 29 April 2010).[7] On the other hand, one of the defendants, Fredrik Neij, stated that 'The Pirate Bay is just a transference service, not a storage service', which would indicate that the individuals maintaining the web site and its functions play a passive role that does not infringe upon copyright law, which is largely designed around the mechanical reproduction of copies (Larsson, 2010; 2013b). This description, then, would correspond to an intermediary service, much like an infrastructure, Internet service provider, or service provider that *solely* mediates, irrespective of what its users use the service for, which is a keystone of the Swedish Act on Electronic Commerce founded in the EU directive on e-commerce (2000/31/EC). Defence lawyer Jonas Nilsson repeated this conceptual link to infrastructure some months later when, after the Supreme Court had denied a review permit, he argued on 'DN debatt' (a Swedish morning newspaper) on 14 May 2012:

> In our view, this is akin to the Post Office being sentenced in court for delivering a letter of illegal contents. Another, and perhaps more relevant analogy, would be for the owners of [C2C sales services web site] 'Blocket' to be convicted because a user had sold a stolen bicycle after placing an advert on their site.[8]

Naturally, in his capacity as defence lawyer, Nilsson wants to impress a rhetorical point here: The logical conclusion to this argument would be his client's acquittal. Nevertheless, my argument, this early in the introduction, is to demonstrate that law

7. SvD NÄRINGSLIV. 29 September, 2010. www.svd.se/naringsliv/digitalt/dag-2-svdse-rapporterade-direkt-fran-the-pirate-bay-rattegangen_5422679.svd

8. DN Debatt. 14 May 2012. http://www.dn.se/debatt/the-pirate-bay-grundare-gar-vidare-till-europadomstolen/ [Last accessed 10 August 2016].

is dependent on *how* we understand events or phenomena which, by no means, are necessarily objective facts. I wish to underline that it is precisely in times of change—such as when digital phenomena emerge in forms we have not previously encountered—that a greater need for negotiations emerges and the conceptual struggle is at its most vigorous. It is then that it is the most political, and perhaps even the most ontologically flexible.

COGNITION, LAW, AND DIGITAL TECHNOLOGY

This book is largely concerned with analysing the impact of technology-related social change. The change we specifically focus on here concerns digitalisation, i.e. there is a clear technological component to the book's theorisation. However, the key point is not digital technology per se, but rather how it impacts the conditions of society, our understanding of this influence, and how it challenges legal rules. In other words, this book develops and analyses the triangular relationship between (digital) technology, law, and cognition (how we understand). The book is mainly based in already published scientific articles that I have prepared for this collection, as noted in the chapter introductions.

The area of *conceptual* metaphors—which has largely inspired my analyses—and what these entail for how we think is a field that is beginning to become relatively well explored, at least with regards to more general and theoretical implications, but less so in relation to the consequences for other disciplines, such as legal or digital technological fields. This aspect of cognitive science research has, to some degree, come to inform legal studies and what it entails for law and legal practice, despite the majority of the studies being based in a North American perspective.

One of the purposes of this book is to develop this a step further and explore the implications in a digital context. There are strong indications that there is a need for this. Digitalisation can be seen as a sort of ontological revolution in which all sorts of events and artefacts are redefined within a shift and interaction between analogue materiality and digital immateriality. In order to develop our understanding of the conceptual significance and the legal challenge that follows from this, a highly applicable, controversial, and much challenged legal area has been selected as a case study: Copyright law.

I theorise around law as a cognitive artefact in general, i.e. how we understand law, using copyright as a specific case study. This focus on the latter object of study is a result of copyright law's relation to digitalisation; copyright in general has so blatantly been challenged by the practices and social norms that have emerged in the wake of digitalisation that it thereby constitutes a strong and particularly important example of conceptual change in relation to technological and social change. One main question therefore becomes: Is there something unique about copyright-related phenomena in digital circumstances that explains this change in our understanding of copyright compared to previous, more physically/analogically mediated circumstances; is there a reason for why copyright specifically is being challenged in digital society? These challenges are also a concern for the copyright-dependent industry which originally molded itself according to conditions that to a large degree no longer apply, and which during the 2000s has, using various methods, attempted to block alternative distribution channels (TPB/Napster), or buy them (Spotify), while simultaneously reinventing and redefining itself to various degrees. The music industry is perhaps the most salient example, followed by the film industry and to some degree the book publishing industry, with regards to copying, handling, and the role of the e-book. Industrial actors often express a somewhat alarmist rhetoric, in which perhaps the most commonly used rhetoric pertains to the artists, authors, and musicians.

OUTLINE OF THE BOOK

While chapter 1 outlines the framework and purpose of this book, chapter 2 develops its theoretical foundation and key concepts. Metaphor theory is developed further in chapter 3, which demonstrates how abstractions in law often receive their meaning from understandings of the 'embodiment' of physical and spatial relations. At this stage, an empirical analysis enters bearing a methodological promise: digitalisation and datafication make various expressions of behaviours and culture traceable and quantifiable, and therefore in some aspects extremely suitable for empirical analysis. In chapter 3 I utilise Google's datafication of a very large set of books in order to study how conceptualisations of law and justice have changed from 1800 to 2000 (in published texts in English). One purpose is to demonstrate the embodiment of law.

Chapter 4 continues to develop the empirical foundation and focuses on the particular case of copyright and file-sharing, which includes a comparative study that shows the difference in how French and American file-sharers conceptualise file-sharing and the purpose and future of copyright. In a sample of approximately 20,000 respondents from the so-called 'Research Bay Study', performed by the Cybernorms research group (which I am part of) at Lund University in Sweden, I demonstrate how French and American file-sharers describe and understand copyright's role and purpose differently. Apparently, even if it is clear that the development of digital technology and smart data transmission protocols is a relatively recent occurrence, there seem to be cultural patterns that echo over presumably much longer time, such as in the difference between the 'intellectual property' of American copyrights discourse and the *droit d'auteur* in the French and continental copyright tradition, developed over centuries.

Chapter 5 is based in one of the models used to calculate the damages in the 2009 Swedish lawsuit against the founders of TPB. Here it becomes clear that there are some inherent anomalies and problems in equating analogical *copies* with digital *copies*. To equate them thusly leads to a conceptual expansion which needs to be problematised, as this chapter does.

After having presented the concept of the copy in the TPB case, I devote chapter 6 to a metaphorical and conceptual analysis of the entire lawsuit, and the struggle to define and label the TPB's entity. Whether TPB was perceived as a platform, a storage or mediating service, or a larger assemblage was crucial in determining whether the key actors had an active or passive relationship to the file-sharing at TPB. I describe the lessons that this lawsuit study teaches us about law's dependence on metaphors, and even more importantly, how the digital world changes and challenges our way of understanding. The preserving force, the conceptual bridge so to speak, is to be found in how we appear to confront and understand new phenomena by using concepts we already bear with us, transferring normativity and framing. In other words, we use old words to describe new things. This creates a sort of lag which further leads us to judge new phenomena by how we are accustomed to understanding other, similar phenomena.

In the context of copyright and legal regulation in a digital age, one could also pose the question: If copyright law in digital contexts is facing as fundamental challenges as many studies indicate, why does the rule remain so obstinately unchanged in its most basic concept? This issue is treated secondarily in both chapter 5 and 6, and in more depth in chapter 7, which analyses both the form and function of (intellectual) property. The chapter deals with copyright regulation meeting the quite rapid societal changes associated with digitalisation, and it does so by reinterpreting Karl Renner's classical texts in the light of contemporary cognitive theory of conceptual metaphors and embodiment. I focus on the notion that the legal norms only appear to be unchanged—the Renner distinction between form and function. This includes social norms, technological development, and changes in social structures in general, which create a social and cognitive reinterpretation of law. I analyse the contemporary push for copyright as property, which I relate to historical claims for copyright as property as well as de facto legal revisions in intellectual property faced with the challenges of digitalisation. Of particular relevance here is what Renner described in terms of property as an 'institution of domination and control', and thus the increased measures for control that are added to a digital context in the name of copyright.

Chapter 8 is primarily devoted to conclusions and revisits the questions that have surfaced in this introductory chapter, and also underlines the role science plays in this changing social landscape. Before we approach that discussion, I wish to first develop more on conceptual metaphor theory, in order to relate this to change—both social and technological.

CONCEPTUAL METAPHORS

Metaphors are a conceptual tool that explain, or provide a way of understanding, various phenomena, behaviour patterns, or events in the world by using more familiar concepts. In brief, the metaphor emerges through the projection of a source domain to a target domain. We experience life as a 'journey' (with a starting point and a destination), and the Internet as a 'web' (where various aspects of different kinds of webs are well suited to be transferred to describe aspects of connected computers and servers).

Often, metaphors are understood as colourful images used to get the actual point across, a sort of rhetorical or poetic tool, and they tend to be regarded as being exclusively linked to linguistic rather than intellective constructions. As a contrast to this minimalistic understanding of metaphors, George Lakoff and Mark Johnson show that metaphors are an inescapable component in how we speak and understand in our daily lives. Our normal conceptual system, they argue, is fundamentally metaphorical by nature. Metaphors and the notions they are founded in arise through our interaction with our physical and social surroundings. Lakoff and Johnson's *Metaphors We Live By* (1980), has come to influence much of current linguistic and cognitive research. The book argues that metaphors play a role in how we think, at a very fundamental level. They exemplify this by noting war metaphors used in argumentation, i.e. argumentation as war: 'Your claims are *indefensible*, he *attacked* all of my arguments, his comment *destroyed* my post, she went *straight for the throat* in the following debate'. The underlying conceptual metaphor or notion controls a rational use of language applied to a certain event, and it does so via the metaphor and its linked system of concepts. In the context of the legal copyright construction, for instance, such underlying notions frame what are considered logical implications of the debated phenomenon. And, whoever manages to control this framing may also control the direction that the debate on copyright develops in. To clarify, there are several ways to understand or theorise around metaphors and their impact on language and thought within cognitive research. The conceptual metaphor theory (CMT) was introduced by the aforementioned Lakoff and Johnson and has since been developed and nuanced, but has also been questioned during the last twenty years. Psycholinguist and cognitive scientist Raymond Gibbs, Jr. recently evaluated the CMT and stated that it, in comparison to basically all other metaphor theories, 'provides important insights into the interaction of embodiment, language, thought, and culture that points to a fuller integration of metaphor studies within cognitive science' (Gibbs, 2011, p. 556). Gibbs explains the empirical advantages of the conceptual metaphor theory (CMT):

> Even if [CMT] does not necessarily account for all aspects of metaphoric thought and language use, this approach has great explanatory power, and must be considered to be foundational for any comprehensive theory of metaphor, as well as for broader theories of human cognition. (Gibbs, 2011, p. 556)

Based on Gibbs arguments for the benefits of CMT, let us then apply the theory empirically to legal and digital phenomena, as well as methodologically to approach an understanding of the relevant conceptual aspects of social change; i.e. how we understand new phenomena, and to what degree we do so via already present or old concepts.

COPYRIGHT AS A CASE

With regards to legal challenges in a digital context, copyright law is an interesting and pedagogical object of study for several reasons. To begin with, copyright is one of the most problematic areas at the intersection of new technologies and law (cf. Lundblad, 2007), which is undeniably indicated by the intensity of the debate from the late 1990s to the present day. Astra Taylor concludes that copyright has become one of the most controversial topics of our time, 'tangled in the contradiction between freedom and restraint, access and control' (2014, p. 150). Copyright is also regarded as an important case at a societal level. For instance, law professor James Boyle (1996) early identified copyright as one of the crucial issues in the construction of the 'information society', but it also signifies a legal development that started long before computers were connected into networks. Boyle argues that, in the last fifty years, copyright has expanded its protection and that this has been done 'almost entirely in the absence of empirical evidence, and without empirical reconsideration to see if our policies were working' (Boyle, 2008, p. 236). Additionally, this 'evidence-free' development runs on 'faith alone' and is a faith that is based on a 'cluster of ideas' that Boyle identifies. This 'cluster of ideas' is of relevance to the underlying conceptions of the copyright debate, which we return to below (cf. Larsson, 2011b), although the 'cluster of ideas' leads to what professor Jessica Litman describes in terms of 'choosing metaphors' in copyright development in her book *Digital Copyright* from 2001. Using this, she outlines 'an evolution in metaphors' that 'conceal an immense sleight of hand', and comments from an American perspective:

> We as a society never actually sat down and discussed in policy terms whether, now that we had grown from a copyright-importing nation to a copyright-exporting nation, we wanted to recreate copyright as a more expansive sort of control. Instead, by changing metaphors, we somehow got snookered into believing that copyright had always been intended to

> offer content owners extensive control, only, before now, we didn't have the means to enforce it. (Litman, 2001, p. 86)

This transition, starting with the conception of mutual benefit for the creator and the public and ending up with the conception of copyright as a system of incentives, completely changes the arguments and rhetoric around it (cf. Larsson, 2011b). This is also supported by law professor William Patry (2009), who has focused on the importance of metaphors in what he describes as 'copyright wars' in *Moral Panics and the Copyright Wars*. The engine driving the given issue of copyright in a digital context runs in the gap between social and legal norms and to what extent the conceptions that construct these dissimilar norms differ. The fact that this regulation is amazingly homogenous across the globe, including Europe, due to international treaties and agreements between states and supranational 'harmonisation' within the EU, makes an analysis of the central metaphors in copyright valid for far more than any single country (Larsson, 2011c). How copyright is conceptualised and how it is formulated will likely affect patterns of creativity, and even how we communicate in digital networks, and it definitely poses questions of privacy in terms of how much of our activities online may be justifiably monitored. This gap, and what is at stake following from it, is what makes metaphor analysis of legal and social norms of copyright both important and attractive (Larsson, 2011b).

The 'gap', however, may be conceptualised in different ways. For instance, Schultz advocates the use of the concept of 'copynorms' to analyse social norms in relation to copyright, as they 'moderate, extend, and undermine the effect of copyright law' (Schultz, 2007). Gracz (2013, p. 39) ties the 'current crisis of copyright law' to technological development and states that the legal system is 'following the wrong pattern' (2013, p. 40).

Strahilevitz (2003a) analyses the influence of social norms in loose-knit groups or in situations where interaction is anonymous. Strahilevitz also analyses the ability of file-sharing software to reinforce descriptive norms in themselves, as it creates the perception that unauthorised file sharing and distribution is a common behaviour, even more prevalent than it actually is (Strahilevitz, 2003b). Strahilevitz made his claim in 2003, and file sharing has undisputedly developed and changed in terms of technology and techniques since then.

The body of empirical studies has seen a steady increase, for example with regards to students (Altschuller and Benbunan-Fich, 2009), and the file-sharing community itself (Andersson Schwarz, 2013; Andersson Schwarz and Larsson, 2014; Svensson et al., 2013, forthcoming) and comparative accounts between countries or cultures (Karaganis, et al., 2012; Larsson et al., 2014a, 2014b). Furthermore, there are now collected volumes that take a broad approach to inherent sociolegal challenges to intellectual property (David and Halbert, 2015) and special sections within the framework of piracy and social change (Burkhart and Andersson Schwarz, 2015). Feldman and Nadler (2006) conducted an experimental study on the influence of law on social norms in the context of file sharing of copyrighted content, which bears a resemblance to a study of norms by Svensson and Larsson (2012). This 'gap problem' of legal norms in relation to social norms can be described as a classic problem in the field of sociology of law, although it has been criticised from time to time (Nelken, 1981; 2009; cf. Gould and Barclay, 2012).

INTENDED AUDIENCE

Is this a book primarily about metaphors, digitalisation, or law? The answer can reasonably be expected to indicate the intended audience. It should be pointed out that this book is not based at the heart of a unilateral scientific discipline; rather, it moves between several disciplines. Its most significant contributions and links are probably to legal doctrine in a broader sense—which include sociology of law—as well as aforementioned cognitive science. It also incorporates a recently emerging field that studies digitalisation and its social impact, which for the sake of simplicity can be referred to as technological and social change, which to some parts resembles what is done in Science and Technology Studies (STS). Certainly, the book is intended for researchers, instructors, and students, but the area of interest is multifaceted. It is not easy to define the relevant academic field here, and, in my view, this may not be an overly important concern. Its main contribution probably lies precisely in being a somewhat unconventional combination of the various disciplines' approach to theory, methodology, and object of study. For cognitive scientists and rhetoricians, the contribution may perhaps lie foremost in the application of already familiar theory in studies on changing digital and legal processes from a conceptual and cognitive perspective. Of course, rhetoricians have concerned themselves with legal processes—particularly in court—for over 2,000 years, and there are a number of cognitive scientific studies with a legal focus. But the overwhelming impact of digitalisation has still been the object of relatively little research, from this collective perspective. It follows, then, that the metaphor-theoretical contribution to a broader legal science might also mean that this book can be of interest for sociologists of law and researchers within law and society. Not the least, the book also provides a fairly new perspective on the challenges that face copyright law in a digital context. Finally, for readers whose interest mainly lies in digital development and the impact of new technologies on society, this book is hopefully of use within a legal focus, but also with regards to a more general need for concepts for new phenomena—and the accompanying processes of how we tend to understand the new through the already familiar and what this leads to in terms of conceptual path dependence, normativity, and lock-in effects.

Just to briefly return to the overarching titles of our digital times, we may conclude that also such a relatively new phenomenon as the Internet and its linked technologies change and mature and so do our conceptualisations of them correspondingly. Interestingly enough, while the Internet used to be an 'information superhighway', a 'cyberspace', or a 'World Wide Web', it has now seemingly transformed into being understood in terms of 'layers', a generator of 'big' data, and preferably consisting of 'open' attributes. While we will return to this development in the concluding chapter 8 below, I will first move on to more thoroughly present the theoretical foundation of this book.

> A world ends when its metaphor has died.
>
> An age becomes an age, all else beside,
> When sensuous poets in their pride invent
> Emblems for the soul's consent
> That speak the meanings men will never know
> But man-imagined images can show:
> It perishes when those images, though seen,
> No longer mean.[9]

9. The poem is part of Archibald MacLeish's 'Hypocrite Auteur' from *Collected Poems 1917–1982* (New York: Houghton Mifflin, 1985).

[2]

METAPHORS AND NORMS

Before addressing any digital phenomena, this chapter will elaborate on the theoretical foundation of the book, which includes the origin and development of conceptual metaphor theory (CMT). It also includes the experiential and embodied aspects of concepts—a central theme in CMT—and the notion of conceptual links in terms of *skeuomorph*—which is not a central theme of CMT. In order to display the importance for law and regulatory development in its relationship to technological development, the terminology of path dependence is utilised. This is a way to emphasise normative linkages—in both design and concepts—over time. Lastly, in order to relate the cognitive and conceptual theory to law as well as to social norms, these concepts are outlined from a sociolegal basis of theory in sociology of law.[1]

Where most metaphor theory focuses the lingual terminology, Lakoff and Johnson's use of CMT emphasises that metaphors connect not just words and phrases, but the underlying ideas or conceptions.

> From the experience of getting sick and lying down, then recovering and arising from the sickbed, we have HEALTHY

1. Some sections of this chapter have been excerpted from or revised based on Stefan Larsson, "Conceptions, Categories, and Embodiment—Why Metaphors Are of Fundamental Importance for Understanding Norms," in *Social and Legal Norms: Towards A Socio-Legal Understanding of Normativity*, ed. Matthias Baier (Farnham, UK: Ashgate).

Conceptions in the Code. Stefan Larsson.

> IS UP/UNHEALTHY IS DOWN, the basis for linguistic expressions like '*high spirits*' and '*feeling low*'. (Ritchie, 2013, p. 70)

A key question lies therefore in how these underlying ideas and conceptions are shaped, and what they are dependent on. Lakoff and Johnson claim that abstract concepts are formed from associations in perceptual experiences, and these concepts provide the basis both for abstract thought and linguistic metaphors. This means that groups of thematically related linguistic metaphors can be used as a tool to detect underlying conceptual metaphors (Ritchie, 2013, p. 85; cf. Larsson, 2011b). Time, for example, is generally conceptualised as if it had physical dimensions, or is located in space. How we talk about time is similar to how we talk about distance and position (Knowles and Moon, 2006; cf. Mattsson and Larsson, 2014).

CONCEPTUAL METAPHOR THEORY

A key idea of metaphors is that they are analogies which allow us to map one experience (the target domain) in the terminology of another experience (the source domain), and thus acquire an understanding of complex topics or new situations. The metaphors often tend to be viewed as exclusively linked to linguistic structures, rather than to thinking and the mind. In contrast to this minimalist conception of metaphors, Lakoff and Johnson showed that the 'metaphor is pervasive in everyday life, not just in language but in thought and action. Our ordinary conceptual system, in terms of which we both think and act, is fundamentally metaphorical in nature' (Lakoff and Johnson, 1980, p. 3). This means that metaphors are not simply a figure of speech, but a 'figure of thought' (Lakoff, 1986), which implies a focus on metaphors not merely as lingual features but as conceptual features. In the words of legal scholar Carl S. Bjerre:

> The cognitivists propose that metaphor, far from being a post hoc embellishment of already-formulated thought, is a tool with which the mind constructs concepts in the first place. Metaphor doesn't just express thoughts in an interesting way; it helps to determine or discover them. Metaphors are a way in which we understand. In a word, they are not just linguistic, but conceptual. (Bjerre, 2005, p. 105)

Further, Lakoff and Johnson argue, metaphors arise from interaction with our physical and social surroundings, in terms of *embodiment* (Lakoff and Johnson, 1999; see also Johnson, 1987; Kövecses, 2008; Lakoff, 1993; Larsson, 2013d; and Winter, 2001, 2007). Bjerre gives a useful example of an embodied metaphor with 'mental capacity': 'This word is literally a reference to containers rather than to mental power, and of course, the mind is not literally a container of any kind' (2005, p. 104). This embodiment, or borrowing from a material reality and bodily and spatial relations, is likely especially visible in digital discourse. For example, when one seeks to explain what The Pirate Bay (TPB) really was, one inevitably used embodied metaphors such as torrents, swarm, domain, and tracker, and this also seems to have been a key argument in the Swedish TPB case with regards to whether TPB was, in fact, a 'bulletin board', a 'platform', or a 'search engine', which was of direct relevance for the outcome in the case (see chapter 6). These conceptual processes more or less are forced to refer to already established concepts, which thereby become normative in controlling the conceptions of the new phenomenon (such as TPB). The underlying conceptual metaphor or idea controls the allegedly rational use of language to describe a particular phenomenon, and it does so via the metaphor and its system for linking concepts. In relation to the construction of copyright law, such underlying conceptions establish a provisional framework for the logical consequences of the phenomenon under debate (Larsson, 2011b). And whoever succeeds in controlling this framework can also manage the course of debates about copyright legislation (Herman, 2008; Yar, 2008).

The conception defines what the socially meaningful use of language for a given phenomenon, and it does so through the metaphor. By emphasising the social aspects of meaning making, context-dependency, and the situated aspects, the sociocultural implications become of great importance. Winter expresses this as, '(m)eaning is a shared social phenomenon that constrains how we as embodied and culturally situated humans understand our world' (2001, p. 315). A further generalisation that we can make regarding metaphor comprehension is that it is mandatory, in the sense that it is an 'automatic' interpretation made by us (Glucksberg, 2008). This means that literal meaning has no priority; the associative paths that create meaning are already present anyway.

CMT is applied and discussed in various scholarly fields and disciplines such as psychology (Moser, 2001); political analysis (Carver and Pikalo, 2008) with contributions, for instance, from Drulák (2008) and Walter and Helmig (2008); sociology of law (Larsson, 2013a, 2013e, 2015) and technology studies (Cass and Lauer, 2004). Of extra importance for this book is the legal analysis that has been conducted based on the work of Lakoff and Johnson—for example, Berger's work on rhetorical choices in Supreme Court decisions (2007), including child custody disputes (2009, see also 2004 and 2011); Johnson's analysis of the misleading 'objectivist' perspective in law (2007); and Winter's

A Clearing in the Forest (2001, cf. 2008 and 2012). In addition, Tsai (2004), for example, describes the historical scepticism toward the role of the metaphor in law. A few scholars have focused on copyright and its enforcement from a conceptual metaphor perspective, such as Herman in an analysis of copyright as property (2008); Loughlan (2006) in an analysis on metaphors in intellectual property; and Yar (2008) regarding rhetorical strategies in intellectual property. In a perhaps less theoretical, but strongly empirical and argumentative way, Patry (2009) too deconstructs copyright law from a metaphorical perspective.

In brief, how conceptual metaphors are constructed discloses *the cultural bias* of our thinking (Kövecses, 2005; Larsson, 2011b, 2012b). If rigid thinking related to metaphors and conceptions is to be understood, it must be elevated to a conscious level where it can be perused and examined. Given that metaphors are so fundamental to thinking and communication, the fact that the same is true of the legal system should come as no surprise. But there are several, very important problems related to the role of law in distributing power and in cases where the legitimacy of the law is at stake.

EMBODIMENT

Embodiment is of great importance for the process of meaning making, according to the focused strand of cognitive science (Gibbs, 2005; Johnson, 1987; Kövecses, 2008; Lakoff, 1987, 1993; Lakoff and Johnson, 1999; Winter, 2001). This follows from the fact that metaphors are such a common and fundamental part and process of language and the mind. This statement means that there is constant borrowing in progress and interdependence with regard to the surrounding context in order for language to become meaningful. In short, metaphorical understanding is often based on our interaction with our physical and social environment. It is derived from bodily sensations and is at its conceptual level, for instance, found in *image-schemas*—e.g. balance keeps you *upright*; more is *up*—derived from adding things to each other and increasing the pile upwards (Lakoff, 1993, p. 240). Conceptual metaphor theories fit well with *the conduit metaphor* system, first identified by Michael Reddy (Reddy, 1979; Larsson, 2013a; Winter, 2007, p. 884, 2001, pp. 52–6). The latter is a systemic set of mappings from the source domain of physical objects to the target domain of mental operations (see Table 2.1).

These are conceptions that also guide and control how new metaphors develop, and can be represented by the mnemonics: IDEAS ARE OBJECTS, KNOWING IS SEEING, WORDS ARE CONTAINERS, COMMUNICATION IS SENDING, UNDERSTANDING IS GRASPING, and THE MIND IS A CONTAINER. The conduit metaphor

Table 2.1 THE CONDUIT METAPHOR SYSTEM

Source Domain		*Target Domain*
PHYSICAL		MENTAL
object	→	ideas
seeing	→	knowing
container (vehicle)	→	words
content	→	ideational content
sending	→	communicating
grasping (receiving)	→	understanding
container	→	mind

system enables us to automatically extend the conceptual mapping by modelling other actions in the physical domain—as a result of embodiment. It is thus meaningful to speak of the 'content' of this chapter, of trying to 'grasp' the concept of norms in sociology of law, or of 'seeing' the point I am trying to make here. This dependency on the physical and spatial can be expressed in the words of Steven L. Winter:

> Thought is not primarily linguistic and propositional, but embodied and imaginative; language is neither entirely arbitrary nor merely socially contingent, but grounded in our embodiment and motivated by our interactions with the physical and social world. (Winter, 2001, p. 47)

We borrow constantly from the physical world, from the objects, the body, and the spatial domain, in order to describe and talk about the abstract. This is where we must be extremely attentive to how our understanding is constructed. Many of the building blocks of this process, even for much less obviously metaphorical concepts, are studied in cognitive linguistics, but also in the symbol- and language-oriented legal (critical) discourse. In terms of the ontological account, the embodiment should, however, probably not be taken as evidence that the values attached to specific metaphors, or any language, can be constructed completely unrestrained by patterns, as a more social constructionist perspective could imply, nor that they are 'natural' and completely rationally fixated (see Larsson, 2012b). There are patterns, and these can be strong patterns which are perceived as 'natural', but they are to some extent negotiated socially and culturally. The point here is not that a freedom exists in general, but that these patterns are probably often different from the rationalistically defined and outspoken patterns and categorisations (Johnson, 2007).

There is research showing a dual connection of metaphor to both body and culture, which we will return to below in chapter 3. Yu (2008), for example, argues that primary metaphors, derived from bodily experience, are likely to be widespread and universal, while complex metaphors, based on conceptual mappings and cultural beliefs, are likely to be more culturally specific.

METAPHORS AND LAW

A consequence of embodiment is that law is also constructed from embodied metaphorical and linguistic operations, relying on the in-the-mind cognitive approach (Johnson, 2007; Larsson, 2012b; Winter, 2001, pp. 22–42). These findings question the 'objectivist' approach in law (Johnson, 2007). As law is an abstract social phenomenon, it is rather the case that the need for metaphors is great. We need to 'reify' and embody law in order to talk about it, think about it, and to enforce it. Winter consequently claims that law can constantly be found to be in a dependent position, which requires object making, a reification, where we talk about law as something physical (2001, 2008). This is particularly noticeable in blatant linguistic metaphors such as: 'the long arm of the law' (Berger, 2009, pp. 262–66) and how we 'take law in our own hands', but perhaps slightly less clear concerning how we 'break' the law, sign a 'binding' contract, or 'trespass' when sharing files that we are not authorised to share. We speak of a *corpus juris*, and without this reification there is no law, which I return to more thoroughly in the following chapter (Larsson, 2011b, pp. 64–65, 2014b; Winter, 2001, p. 334).

There is a growing field of research on the interpretation of legal metaphors. Many people with an interest in legal analysis and influenced by this school of metaphor theory start their presentation with the conflicting aspect of metaphors in law. They often do that by citing the early American legal realist Justice Cardozo, who observed that 'metaphors in law are to be narrowly watched, for though starting as devices to liberate thought, they end often by enslaving it' (see Berger, 2004; Herman, 2008; Patry, 2009; Winter, 2008). Ironically enough, and as is often pointed out, this statement by Cardozo uses at least two important metaphors—liberation and slavery (Loughlan, 2006). Cardozo's view follows in the same tradition as the famous enlightenment philosopher John Locke, who condemned the metaphor and other figurative speech as 'perfect cheat' and insisted on literal prose, 'if we speak of Things as they are' (Locke, 1975 (1690)). Conceptual metaphor theories collide with a more formal approach to meaning and truth. A formal approach underlines the objective, literary, and linear thinking where metaphors are considered merely colourful figurative linguistic symbols or expressions, used to express a point clearly. This includes law and legal thinking. Robert Tsai writes:

> Legal scholars have traditionally understood metaphor as, at worst, a perversion of the law, and at best, as a necessary but temporary placeholder for more fully developed lines of argument. On this view, metaphors are vague and inherently manipulable, appealing to base instincts, whereas explicit legal argumentation represents the rigorous, authentic core of law. (Tsai, 2004, p. 186)

This conception in the legal field has been durable and strong. The British philosopher Jeremy Bentham believed that metaphors were the antithesis to legal reasoning (Bentham and Ogden, 1931). The traditional legal perception of the unimportant place and function of metaphors in law is, in other words, completely at odds with what language and cognition research has shown from 1980 onwards. Philosophy professor and cognitive scientist Mark Johnson uses these insights to study the language-based judicial system and to ask how we can achieve a better understanding of the way legal knowledge and reasoning works, and sees part of the answer in cognitive legal research, in a 'cognitive science of law', i.e. how thinking works in relation to legal concepts and legal decision making (Johnson, 2007). Johnson argues that legal reasoning and legal concepts are based on a sort of self-image of objectivity. This 'objectivist view' includes the idea that concepts have strict limits and are defined by necessary and sufficient conditions, which contradicts the empirical findings in cognitive and linguistic research (see, for instance, Winter, 2001). Johnson's main point in including cognitive science in the study of law is that the legal 'objectivist view' is based on an incorrect understanding of how thinking and language works (Johnson, 2007, p. 847).

As mentioned, there is a growing body of literature that deals with the analysis of law based on the work of Lakoff and Johnson (see Berger, 2004, 2007, 2009; Blavin and Cohen, 2002; Bröchner et al. 2015; Herman, 2008; Hunter, 2003; Johnson, 2007; Joo, 2001; Larsson, 2011b, 2012b; Morra, 2010; Ritchie, 2007; Tsai, 2004; Winter, 2001, 2007, 2008). An example on how metaphor studies have found their way into policy research and political analysis is the above mentioned collection of chapters in Carver and Pikalo's *Political Language and Metaphor* (eds., 2008). The anthology emphasises the impact of our metaphoric language, and puts forward the productive element of metaphors in the subtitle, 'interpreting *and* changing the world'. However, many of the contributors draw their theoretical foundation from the CMT of Lakoff and Johnson, and inspiration is also found in Black (1962, 1979) and the pragmatic philosophy of Schön (1979), for instance.

Metaphors cannot be studied solely in relation to legal norms, but must arguably be studied in relation to social norms as well. It has been suggested that studying metaphors may also be fruitful when examining the imperative essence of social norms (Larsson, 2011b, pp. 52–3, 65–8, 123–4, 2013a, 2013e), as defined by Svensson (2008, 2013) and Hydén and Svensson (2008). The major methodological difference between studying legal metaphors versus socially embedded metaphors that affect social norms is found in the formalised character of the legal norm. The legal formalisation creates certain and fixed metaphors, and hence 'locked-in' conceptions, whereas socially embedded metaphors and conceptions probably require another type of empirical evidence when studied (Larsson, 2011b, pp. 85–7).

It has been shown that legal metaphors can allow rather substantial changes in meaning without necessarily changing concepts, as I have elsewhere (Larsson, 2011b) shown in relation to 'intellectual property' and the 'copy' in copyright law in a digital society (Larsson 2013b, 2014c), which we return to in chapter 5. This means that the study of legal metaphors is a study of fixed metaphors, whereas the study of the metaphors that control a social norm needs to be separately formulated. This is the case especially in circumstances where a study is conducted of social norms that do not correspond to legal norms—that is, that are not based in a preformulated conception of a norm.

COGNITION AND NORMS

Part of the claim this book makes lies in the possibility of combining CMT and this type of cognitive research with sociolegal theory in research on the relationship between law and society. This means, in that case, highlighting the cognitive aspects of the formation of normative structures or, more simply put, that how we think we ought to behave and regulate is dependent on how we understand (the phenomena of) the world. The legal case utilised in the book is copyright, placed in the context of, and challenged by, digitalisation. For a theoretical marriage to be possible between sociolegal science and CMT, knowledge of the link between how concepts are shaped and worldly phenomena needs to be reinforced. Based in a more traditional concept of metaphors, this link is not necessary or stressed, whereas in the CMT proposed by Lakoff and Johnson, we constantly see an emphasis on the embodiment of concepts—that is, the dependence on body and spatial dimensions for the shaping of abstract thought. Furthermore, the emphasis in CMT on the metaphor as essential for thought points out what Gibbs (2014, p. 30) calls ‘the social nature of ordinary thinking’, which has not been at the core of this theoretical development, but is an element I focus here. Gibbs claims that it may be impossible to disentangle private metaphoric thought from metaphoric social action:

> Even if metaphoric concepts are essential for certain aspects of abstract thought, they typically emerge in specific contexts in which people attempt to better coordinate their own actions and their interactions with other persons. In this way, conceptual metaphors are both cognitive, in the sense of shaping thinking, and social, in the sense of facilitating people’s social behaviors. (Gibbs, 2014, p. 30)

Gibbs argues that conceptual metaphors, being inherently situated, must be studied in terms of the dynamic couplings between language and social actions, not just as pure cognitive entities passively stored in individual minds (Gibbs, 2014, p. 36). Gibbs, in other words, offers a link from cognitive theory to social theory, which is of relevance to both strands. Somewhat similarly, an important point of most sociolegal research is law's *social dependence* (Aubert 1972, p. 13; Mathiesen 2005; Hydén 1978, p. 26; Svensson and Larsson, 2012). This is also one of the key arguments put forward for using the concept of norms to analyse society (Hydén and Svensson, 2008; Svensson, 2008, 2013).

Emphasis is here put on the importance of understanding how language, meaning, and thought are connected. A main advantage of the metaphor-theoretical connection to general norm analysis lies in how language-based expressions and metaphors relate to the mind, and how phenomena are framed and are thereby both controlled and enabled by how different conceptions construct lingual metaphors. This applies to the most mundane and everyday events as well as to law. This perspective indicates the importance of detailed studies of the surface structures, which have the potential to reveal underlying conceptions that may control a legal or social norm (Larsson, 2011b; Larsson and Hydén, 2010).

According to Johnson, it is in a Western moral tradition we make ethical decisions by applying universal laws to concrete situations. Contrary to this conception, Johnson argues that research in cognitive science undermines this view of normatively controlled and enlightened behaviour and reveals that imagination has an essential role in ethical deliberation (1993). How we conceptualise and imagine how we can act not only has implications for how we actually act, but also for what we think is just and appropriate behaviour. These imaginative paths are probably not unforeseeable and random, as Winter puts it:

> Despite the fact that it is conventional to think of imagination as random, unpredictable, or indeterminate, it is actually orderly and systematic in operation. (Winter, 2001, p. 259)

Arguably, it is the task of the sociologist of law, among others, to depict the implications of these systematic and orderly processes of imagination in terms of their impact on behaviour and social norms. However, for us to understand these implications, we must turn to how metaphors operate in detail, as well as relate them to underlying conceptions and how conceptual mappings work.

American professor of law Linda Berger has successfully used metaphor theory to analyse judicial decision making in custody disputes (2009), rhetorical choices in (U.S.) Supreme Court Decisions on campaign finance regulation (2007), and to help lawyers shape the law (2004). One of her conclusions is that lawmakers cannot avoid being affected in their decision making by myths, metaphors, and symbols. Berger analyses cases that involve determining which parent is best suited to have custody, when there is often no rational basis for choosing one parent over another. Decisions must, however, be taken, and Berger argues that there are 'embedded knowledge structures' that influence judicial decision making in a negative and unconscious way. This is sometimes a disadvantage to individual families that do not conform to a more traditionally rooted picture of what families ('should') look like. Berger makes what she calls a rhetorical analysis in order to 'uncover the symbols and stories that affect judicial decision making and then to construct arguments that may overcome deeply rooted constraint, help individual clients, and persuade policymakers' (Berger, 2009, p. 260):

> The rhetorical analysis indicates that the best interests of the child standard fails to explain child custody outcomes, and the analysis suggests that the cognitive setting for custody disputes—cluttered with outmoded metaphors, simplistic images, and unexamined narratives—interferes with the

> ability of judges to attend to complex and radical transformations of parent-child relationships. (Berger, 2009, p. 260)

In this sense, she argues that a better understanding of the cognitive role of the metaphor can help display how law is dependent on symbols and metaphors, which is of extra importance when these become 'outmoded' or perhaps out of step with those of society in general.

Categorisation and law

Categorisation is a fundamental activity in language, thought, decision making—and not the least law—which probably also correlates with how norms affect behaviour. How we categorise—be it socially reproduced or neurologically connected—is likely relevant for determining how social norms are alerted in a given situation, and undeniably which legal norm should apply to each specific case. The classical Aristotelian approach to categorisation sees it as based on grouping objects according to their similar properties. According to the classical view, categories should be clearly defined, mutually exclusive and collectively exhaustive (Lakoff and Johnson, 1999, pp. 373–414). However, a cognitive approach following Lakoff and Johnson's theories (1999) renounces the classical approach in that it accepts that natural categories are graded (they tend to be indistinct at their boundaries) and inconsistent in the status of their constituent members. It is not that the classical view is entirely wrong, according to the cognitive approach; it is just that the categorisation based on shared properties only displays a (small) part of the story (Lakoff, 1987, p. 5). This means that even categories are to some extent culturally biased and connected to embodiment. This has implications for the study of norms in the sense that whenever it concerns categories, the study should include an empirical approach to the categories the norms are dependent on, instead of an ideal or analytic definition of the category. How we categorise (in practice) probably has implications for how we reason about and justify behaviour.[2] In a study on how online file-sharers justified their likely often illegal behaviour, Andersson Schwarz and I (2014) show that how actions are perceived—categorised—is of relevance for how they are justified. From an empirical account, it became clear that there were different conceptions of online file-sharing at play. The justifications were analysed within a range of 'it's unstoppable' to 'culture in general doesn't suffer' and 'it's democratic' (Andersson Schwarz and Larsson, 2014, pp. 223–24) and many additional categories were defined on the basis of the respondents' answers. In brief, how the role and effects of online file-sharing were conceptualised—a process in which metaphors played an important role—affected the respondents' normative justifications.

In a more neurological account covering the same legal field, Goodenough and Becker (2008) studied aspects of how people experience property in relation to intellectual property by looking at neurological functions. They concluded that traditional approaches to property, in particular those relating to tangible objects, are supported, at least in part, by functional neurological structures or networks in the brain that help to recognise property-related choices and to link them to emotional, inhibitory processing. Further, they state that these property-related primitives and/or networks are not readily elicited and mobilised by concepts of intellectual property. Following the point

2. See, for example, Knutagård (2009) regarding the role of categorisation in the case of a Swedish municipality's organisation of its work for homeless people, and how categories are used to legitimise and explain homelessness.

made by Goodenough and Becker, this could as a rough general point be expressed as how we experience reality is relevant for how we define what we find right and wrong. The Goodenough and Decker conclusion, in the field of 'cognitive jurisprudence', suggests that people do not experience illegal file-sharing as stealing in its physical sense. This is supported by studies which have found the social norm corresponding to copyright to be extremely weak (Svensson and Larsson, 2012; see also Larsson et al., 2012b)—regardless of the letter of the law.

Additionally, if we focus law itself, a common practice in law-making and legal decision making is to form categories. The legal categorisation is often viewed from a conventionalist standard viewpoint, which states that categories are descriptive, definitional, and have exact delimitations. It follows from this view that any example chosen from within a category will represent the category equally well. In contrast to this view, cognitive science claims that there is a pattern, and it is empirically detectable, but it is often not the same pattern as the conventionalist or rationalist view. From this practical approach to categorisation follows that certain members of a category are generally regarded as more representative members of that category than others. For instance, when speaking of furniture, *chair* is more frequently cited than *stool* (Rosch, 1975, 1978; see for instance also Lakoff, 1987). This example of *prototype theory* is a departure from Aristotelian logic—that categories are logical, clearly bounded entities, whose membership is defined by an item's possession of a simple set of criterial features, in which all instances possessing the criterial attributes have a full and equal degree of membership. Prototype effects mean a sort of graded categorisation, again as opposed to a traditional objectivist view (Amsterdam and Bruner, 2000; Lakoff, 1987; Winter, 2001). In addition to this, the categorisation that we perform in our everyday lives, just as with the use of metaphors, is automatic and unconscious (Lakoff, 1987). This is a fact that makes any study of norms, which includes categorisation of some sort, in need of at least some reflection upon how this categorisation is carried out.

SKEUOMORPHS AND THE CONCEPTUALISATION OF THE DIGITAL WORLD

Metaphor research has yet to embrace much of the digital revolution and what it means for how we shape concepts, to what extent there are—near—universal elements, and to what extent there are local or cultural diversities when creating meaning around similar digital artefacts. Ritchie (2013, p. 79) mentions that '[t]he Internet has yielded a raft of culturally based metaphors, although given worldwide adoption of computer technology these are likely to be quite widespread', but does not emphasise a culturally differentiated conceptualisation of digital technology and phenomena.

Wilken (2013) compares and contrasts Internet-based metaphors against those in the mobile literature, as part of a 'new media discourse', to find implications of these metaphor selections for our conception of and engagement with mobiles. He concludes that 'metaphor is inescapable, all pervasive, and the study of it is important if we are to grasp fully the way we experience and make sense of the world, especially that which we take to be "new", novel, or which evades apprehension (the seemingly "unknown")' (Wilken, 2013, p. 633). Sandra Greiffenstern (2010) studies the lingual effects of computers, the Internet, and computer-mediated communication on everyday English.

To begin with, the Internet includes an English bias, leading to many non-English natives adapting to an online set of English concepts and phrases, even if the non-English web is increasing strongly as of lately (Greiffenstern, 2010, p. 4). Another question of great interest that Greiffenstern also asks is how the increasing use of technology change how we perceive the world around us. Even if one could question the increase in use of technology, it sounds (a bit too) deterministic and is a common misconception, or perhaps just an inexact use of words, that tends to equate 'technology' with 'digital technology', especially computers, including smartphones. One should not forget that the pen also constitutes (writing) technology, and the paper with its written letters a sort of external memory. Nevertheless, Greiffenstern states that (digital) 'technology has a great impact on how our view of the world has changed. But at the same time, we mainly use already known concepts and phenomena to describe and understand new technology' (2010, p. 6; cf. Larsson, 2015). This adaptation is, of course, of great interest when it comes to regulation and behaviour related to new technology.

Concepts are constantly transferred to new phenomena that carry similar elements and I utilise the concept of the *skeuomorph* to analyse and discuss that process in relation to new technologies and digitisation. The development of information and communication technologies, combined with their massive distribution and use, has created a considerable need for labels and concepts that can describe the multitude of phenomena that follow (Cass and Lauer, 2004, p. 255; Larsson, 2013c, 2015). Even when the phenomena in their technical nature could be argued to be brand new, concepts for preexisting phenomena are metaphorically transferred, because they share some similar elements or possible associative paths; for example 'computer virus' or 'cyber-trespass' (Klang, 2003; Wong, 2006). Some features from the previous phenomenon fit well, while others do not. A skeuomorph provides us with familiar cues in an unfamiliar domain by presenting parts that make new things appear old and familiar (Gessler, 1998). The re-use, or extended use, of a metaphor is often quite necessary and 'natural':

> When the technological media of an artifact changes, some characteristics of the previous media are left behind, others are brought forward intact into the new media, while still others are brought forward in a modified form. In the transition between the non-digital and digital media, a learning process occurs where users employ metaphors from the non-digital representation and process to orient themselves to the novelty of the new media. (Cass and Lauer, p. 255)

This is to emphasise how familiar and 'non-digital representations' take part in a process of conceptual change that is at least to some extent deceptive, and to some extent likely to be qualitatively informative. The 'skeuomorph' is a concept that can be used to describe precisely that process (Larsson, 2011b, pp. 62–3, 101–02). Consider, for instance, the examples of transition from regular mail to e-mail or from analogue photography to digital imagery (Cass and Lauer, 2004). The skeuomorph means a type of reuse of information, but a reuse that comes with benefits as well as pitfalls.

Recognition is created through this use of older concepts, but it is potentially a deceptive recognition. It is important to notice that the skeuomorph is not entirely a physical entity; the reason it can create this recognition in the first place is due to the fact that it collaborates with our cognitive images, our conceptions, and how we understand reality. In the literature, the skeuomorph is a type of metaphor that is often represented by an object. Some speak of skeuomorphs as 'material metaphors', since the term is often used to refer to the design of artefacts that bear some resemblance to older and established artefacts (Gessler, 1998). Cass and Lauer, on the one hand, emphasise physicality and materiality in the term, but on the other describe its direct link to cognition:

> People use physical (skeuomorph) and conceptual metaphors to orient themselves with new technology by understanding new functions in terms of earlier technological versions. Since new technology is adopted at varying rates and varying times, multiple versions exist at any given time. (Cass and Lauer, 2004, p. 252)

This is why skeuomorphs cannot only be described as physical entities. It is in the interpretation that the important aspects lie. This is also why Cass and Lauer extend their definition:

> Skeuomorphs, in addition to being physical features, can also take the form of ideas or metaphors. When new artifacts are presented to the public, many times they are described with metaphoric allusions that are grounded in prior iterations of that artifact. These metaphors assist people in their transition to understand and use new technological processes and artifacts. When the new artifact is described using terminology from a prior iteration, this influences one's intent and encounter with the new artifact. The new artifact is initially understood using the norms and interpretive scheme of the old artifact, which aids in both transition to and adoption of the new medium. After collective learning about the new artifact that occurs, artifact users may discover new functionalities beyond what the 'transitional functions' suggest

> and places where the metaphor breaks down. (Cass and Lauer 2004, p. 255)

Note that it has implications for norms too, according to Cass and Lauer. By borrowing an understanding of a phenomenon from an older concept, the norms connected to that older concept will likely connect also to the new phenomenon. A main interest here is how this dialectic between the old and the new is relevant not only for how we metaphorically interpret aspects of our reality, but also for how this process sustains whatever normativity may accompany the first object or entity and is transferred to the latter, where 'the new artifact is initially understood using the norms and interpretive scheme of the old artifact' (Cass and Lauer 2004, p. 255). This, I argue, is of clear relevance also for legal developments connected to technological evolution, including digital phenomena such as TPB—which I focus on in chapters 5 and 6.

For example, when we conceptualise TPB, we are inevitably surrendering to a conceptual reuse that is massive: It is found in the 'domain' name, relies on 'torrents' to be found by a search 'engine', takes place in a 'swarm', and has moved toward using 'magnet' 'links'. It is important to remember that this is not something strange or an abnormality. It depicts how conceptual developments happen, in the sense that 'abstract thought is largely metaphorical' (Lakoff and Johnson, 1999, p. 3), and therefore a natural part of conceptualisation. This is arguably particularly evident in times when new types of phenomena emerge more frequently, or a new technology has an uncommonly great impact on society.

One could even argue that skeuomorphism is central or integral to media's own development, that this 're-mediation' is necessary for it. The mediation is not only an assemblage or ecology at any given point in time, but also linked to passed forms of media. Bolter and Grusin (1999) write that, 'each act of mediation depends on other acts of mediation' and that 'media need each other in order to function as media' (1999, p. 55).

SKEUOMORPHS AND CONCEPTUAL PATH DEPENDENCE

My point here is to emphasise how already familiar non-digital images become part of a conceptual change that highlight some aspect while hiding others. The concept of the skeuomorph helps describe this process (Larsson 2011b, pp. 62–3, 101–02, 2013e). The skeuomorph is, in other words, a sort of recycling of information, while simultaneously being both an advantage and a disadvantage. Literature critic and researcher N. Katherine Hayles, for example, emphasises the duplicity of the skeuomorph:

> The new becomes more acceptable when it refers back to the earlier iteration that it is displacing, while the earlier iteration becomes more valuable when it is placed in a context where we can experience the new. A skeuomorph simultaneously focuses on the past and future, while reinforcing and undermining both. (Hayles 1999, p. 17)

By using old concepts, familiarity is created for new phenomena. This familiarity is possible precisely because it interacts with the cognitive images and concepts that we use to understand our reality. As mentioned, this conceptual bridge has normative implications. By conceptually linking new and old phenomena, norms linked to the former will also be transferred and become linked to the latter. Part of what makes the skeuomorphs interesting in relationship to the argument of this book lies in just how the link between how old and new phenomena includes normativity. There may be a *conceptual path dependence* that locks in a particular understanding of a phenomenon that *could be* understood, and regulated, in a different way.

When it comes to 'path dependence', there are multiple fields where this theoretical figure has had particular explanatory relevance, such as in spatial planning and in technological design. I focus primarily on what I call conceptual path dependence, i.e. when different ways of understanding certain phenomena become subject to 'lock-in effects'. It is related to time and process, as clearly emphasised in Richard Posner's article on law and its 'past-dependency', where law is found to be 'the most historically oriented' and 'backward-looking' of the professions (2000, p. 573; see also Hathaway, 2001). The theoretical concept of path dependence is also used both in historical institutionalism (Pierson, 2000; Pierson and Skocpol, 2002) as well as in historical sociology (Mahoney, 2000). Its origin is mostly referred to as to be found in the study of technological development to explain how less optimal technologies survive at a later stage that offers better solutions to a given need or problem.

Path dependence theory has been used to describe the fact that political choices made when an institution is formed, or when a policy is initiated, will have a continuing influence (Peters, 2005, p. 71). Although much of the literature on lock-in effects and path dependence has been applied to explain technological development, it has also been applied to other processes, including legal (Gillette, 1998; Hathaway, 2001; Larsson, 2011a, 2011c). Regulations can provide analogies to technical standards (Adams, 1996). Furthermore, the idea of path dependence is closely linked to *transaction costs*, which aims to explain the lock-in effects in terms of threshold-effects for deviations from 'the path'.

SUMMING UP

Language is fundamentally metaphoric in nature and the framing aspects of our perhaps surprisingly metaphoric language is of great relevance, not only to legal analysis in a manner not generally spoken of or acknowledged, but also to the analysis of norms other than legal norms. Metaphors in language, and law, do not simply passively describe reality for better or worse. In fact, quite the opposite occurs: Metaphors in a sense produce (our construction of) reality, since they frame our conceptions in certain patterns, and categorise concepts into seemingly necessary, but often falsely indispensable, categories. Much of this stems from the fact that our conceptual system is fundamentally metaphoric in nature, and thus relies on embodiment and the surrounding physical and cultural context. Furthermore, although conceptual metaphor analysis is thankfully applied to legal norms due to their often text-based sources of interpretation, this type of analysis may also aid in understanding the operation of other types of norms as well. For instance, a metaphor representing a particular social norm may reveal underlying conceptions that affect and are relevant for a particular behaviour, without being spoken of or being consciously acknowledged. The analysis of metaphoric patterns is a method to outline what the underlying and governing conceptions are. The theory's explanatory function can further be used to understand conceptual change, such as the particular linking between an old and a new and somewhat similar artefact or phenomenon, a skeuomorph.

The definition of a metaphor is to some extent similar to that of the skeuomorph in that they both carry meaning from one domain to another, but there are at least two nuances of important difference to point out. First, skeumorphs are explicitly used in (interaction) design—particularly so for user interfaces (UI), as well as archaeology—as an accepted concept, which focuses symbolic, functional, and aesthetic elements more than words and lingual expressions (Gross et al., 2014). Secondly, where the metaphor concept tends to focus the meaning making at the target domain, skeuomorphism emphasises the link between the new phenomenon and what is often described as an historical artefact or design. It is this link that carries values, norms, and expectations from the former iteration to the latter, making it easier—theoretically—to connect notions of transferred normativity to the labelling of the new.

An elaborated account of the underlying conceptions can help to deconstruct the frameworks or principal cognitive models on which a norm is based, or to which it correlates. A part of the necessity of doing so lies in the fact that we may not even be aware that a norm is based on a particular conception in the first place; we might perceive it as 'objectively' describing the circumstances when, in fact, it is framed within a particular conception, and perhaps bears a cultural or technological bias. For example, metaphor analysis can both reveal hidden and automatically associated values in legal metaphors, as well as outline the conception on which a particular law or legal rule is based (and framed).

Recognition of metaphorical thought, and the methods of conceptual analysis, demonstrate how legislative statutes express significant aspects of our social reality which cannot be devalued by reductive approaches to legal reasoning (see Larsson, 2011b; Winter, 2008). A main point is that, although meaning is very much bound to specific patterns, these patterns can be, and probably often are, different from the 'objectively' defined patterns of meaning. In short, the approach of cognitive linguistics and CMT offers a toolbox of deconstructive tools for norm analysis, which I will utilise in the following chapters, both empirically and analytically.

[3]

THE EMBODIED LAW

'For sin shall not have dominion over you: for ye are not under the law, but under grace'.

Romans 6:14.[1]

This chapter[2] uses embodiment and the experiential basis of conceptual metaphor to argue for the metaphorical essence of abstract legal thought. Abstract concepts like 'law' and 'justice' need to borrow from a spatial, bodily, or physical prototype in order to be conceptualised; this is seen, for example, in the fact that justice preferably is found 'under' law. Three embodied categories of how law is understood are examined in this chapter: *law as an object, law as a vertical relation*, and *law as an area*. The Google Ngram Viewer, based on the massive library of books that Google has scanned, has been used to study legally relevant expressions over time within each of these three categories, from 1800 to 2000. In addition, the chapter suggests a type of analytical method of 'metaphor triangulation'—that is, the replacement of prevailing lingual metaphors with unusual ones in order to

1. King James Bible (Cambridge ed.).

2. An earlier version of this chapter can be found in the article published in the *International Journal for the Semiotics of Law* 27, 613–626 under the name 'Justice "Under" Law—The Bodily Incarnation of Legal Conceptions Over Time' (Larsson, 2014b).

increase the level of awareness of what conceptual content the prevailing metaphors involve.

CORPUS JURIS

The intertwined status of law, language, and meaning is often stated but not always in detail clarified (Engberg and Wølch Rasmussen, 2010). If you consider the implications of time in relation to law and legal meaning, you have to consider the implications of change in the interpretation of the explicit concepts. This may indicate that the way we conceptualise law is a culturally bound pattern that shifts in the same sense society shifts and its different institutions with regard to some parts remain rather stable and to some others undergo tremendous change. In chapter 7 I deal with how the meaning of property is facing a tremendous reinterpretation in a digital context, analysed through a theoretical combination of the sociolegal classic by Karl Renner and the more modern findings regarding CMT. This would assert that even when the form, the explicit wording, in law is constant, it's meaning and interpretation may still be under radical change. Hutton (2009) demonstrates the connection between meaning, time, and the law, and discusses the 'plain language' movement in law that is very much criticised from the aspects of conceptual and temporal transition. The notion of plain language is premised on the existence of a 'stable set of signifier-signified relations' (Hutton, 2009, p. 280), which 'represents a claim that law can be a self-sustaining, self-justifying and self-explanatory form of reasoning, that law can exhibit fully its own mechanisms and rationale' (Hutton, 2009, p. 280). A time- and culture-dependent view on conceptions of law would refute this 'plain language' understanding of law.

Seeing the conceptualisation of law as something embodied, in the sense that conceptual metaphor theorists speak of embodiment, would be one such platform for critique.

To be more obvious: We can *break* the law. Law can be enforced *upon* us. Certain individuals or corporations may seem *above* the law. The borrowing of physical or spatial concepts—often referred to as embodiment—seems important for how we understand and talk about law. But, to what extent are the conceptualisations of law changing throughout history? And how can we measure or study it? By drawing on a growing literature on CMT and embodiment and what this means for law, mainly accounted for in the previous chapter, this chapter focuses the metaphoricity of how we generally conceptualise justice and law. This means that the chapter traces a development within cognitive linguistics that regards CMT, categorisation and—what is of key relevance here—embodiment, and comments on some of the consequences this has to law, legal processes, and legal analysis. For example, Lakoff and Johnson regard spatial relations to be 'the heart' of our conceptual system (1999, pp. 30–1). And we use spatial-relations concepts unconsciously as we impose them via our perceptual and conceptual systems: '[w]e just automatically and unconsciously 'perceive' one entity as *in, on* or *across from* another entity' (1999, p. 31; cf. Glucksberg, 2008). These spatial-relations can either be rather universal and in some cases culture-specific. For example, the Hawaiian language offers a distinct concept of spatial-relations that completely makes sense in relation to the particular topography of the mountainous Hawaiian Islands: *Mauka,* which means towards the mountains. On an island that is pretty much circular and always has a mountain in the middle, the direction creates a meaningful concept. This concept would not be meaningful in Southern Sweden, for example, because there are no mountains there. In this sense, many concepts are directly dependent on the embodiment of a surrounding context. The main point here is that we rather inevitably borrow conceptualisations from physical and spatial phenomena in order to understand abstract concepts, such as 'law' and 'justice', and they can be both (near) universal or culture-specific (Kövecses, 2005).

The main argument for the chapter, in brief, is that *we cannot think of justice without embodiment; we cannot understand law without metaphor.* This would then likely mean that there is no 'pure' comprehension of, for example, 'justice' and 'law'; there is no true objectivity to speak of. These abstract concepts need to borrow from a spatial, bodily, or physical prototype in order to be conceptualised, which may have a temporal, cultural, or technological dependency. This chapter addresses this fact and discusses what it means in terms of freedom of tampering with the conceptualisations, i.e. to what extent it really is a freedom or just the clarification of other patterns of meaning-making than those traditionally put forward in an objectivist paradigm. May we, for example, understand more of a given abstract concept if we consciously use uncommon metaphors to conceptualise it as a type of 'metaphor triangulation'? As mentioned, a few examples have been identified here, to elaborate on the main argument more clearly: the conceptual binding of law and justice to a notion of spatial relations such as 1) *law as verticality* (under/above), 2) *law as an object* (and body), and 3) *law as an area*. These are studied through an analysis enabled by the Google Ngram Viewer, which is a database consisting of over five million books, scanned and indexed by Google. The time span chosen for this study regards books published in English between 1800 and 2000. The attempt in this chapter is to be methodologically and empirically probing and tentative when it comes to an understanding of law over time.

EMBODIMENT AND THE CREATION OF MEANING

As outlined in chapter 2 above, a key point in the development of CMT set off by Lakoff and Johnson in 1980 is that that we constantly borrow from one source domain in order to make sense of abstract phenomenon as a target domain. They, and several followers, connect this closely to the experiential processes of embodiment— that is, that these metaphors often have an origin in an understanding of physical entities of spatial relations (see subchapter 2.2 and Gibbs, 2005; Johnson, 1987; and Larsson, 2013c). The term 'embodiment' does not only mean the borrowing from the notion of physical bodies to describe abstract thought but also space and spatial relations; that being said, it is rather common to focus the bodily aspects. Significant for embodiment is that a metaphor can serve as 'a vehicle for understanding a concept only by virtue of its experiential basis' (Lakoff and Johnson, 1980, p. 18). There needs to be some kind of experience that guides the way in this abstraction, which means that embodiment comes quite naturally as filling a need we have (even this very point, in being abstract, needs metaphors to be expressed—'metaphors as vehicles'). In fact, the vehicle metaphor—for metaphors—is also well suited considering its Greek origin: *meta pherein* literally means 'to carry over' or 'to transfer'. Steven Winter expresses the experiential basis in terms of recognising patterns:

> The very capacity of the brain to recognize patterns and form concepts depends on these structures of bodily experience. These basic structures or image-schemas—such as BALANCE, PART-WHOLE, OBJECT, SOURCE-PATH-GOAL, FORCE-BARRIER,

> and CONTAINER—provide structure to human thought and a measure of apparent unity and determinacy in our interaction with the world. (Winter, 2001, p. 15)

Winter explains that these 'image-schemas' are 'neither representations nor literal "pictures"' but 'schematics that emerge from cross-modal linkages in neural processes that transcend any specific sensory modality' (Winter, 2001, p. 15). So there are some sorts of linkages between the experiences from the body that aid in sorting amongst the abstractions. Both Winter as well as Lakoff and Johnson stress the fact that embodied metaphors are neurally structured—that is, the neurological system is wired to form meaning-making patterns from emergent spatio-temporal and kinaesthetic experience. I will further outline and discuss the implications of this below, including the complexity this forms when compared to the social and cultural impact on conceptual metaphors.

It is in this 'experiential' focus that Lakoff and Johnson continued much of their pursuit to understand the conceptual and cognitive processes of the mind in relation to language. Johnson has addressed the bodily basis for meaning-making in language and what this means for reason (Johnson, 1987), and the practices of how morality functions in imaginative patterns (Johnson, 1993). A central claim in Lakoff and Johnson's *Philosophy in the Flesh* (1999) is that reason is, in practice, differently grounded to what traditional western thought has assumed. This has bearing not only on law and legal reasoning but also on moral reasoning: according to Johnson (1993), the Western moral tradition states that we make ethical decisions by applying universal laws to concrete situations. Contrary to this conception, Johnson argues for how research in cognitive science undermines this view of normatively controlled and enlightened behaviour and reveals that imagination has an essential role in ethical deliberation (Johnson, 1993; with regards to social norms, see Larsson, 2011b, 2013a).

Cultural contextuality

With embodiment follows an acknowledgement of the dependence on context for how concepts receive meaning. This opens up an enquiry as to what extent there may be more universally functional meanings and to what extent this process for an explicit conception is culture-specific. For an in-depth account on differences between universal and culturally based variation in metaphor, see Kövecses (2005). Kövecses addresses how the variations occur, and states that 'universal physiological features provide only a potential basis for metaphorical conceptualization' (2005, p. 248). Winter also acknowledges the cultural dimension: '[m]eaning is a shared social phenomenon that constrains how we as embodied and culturally situated humans understand our world' (Winter, 2001, p. 315). Another way to express this is from the opposing angle, in the sense that 'the particular human embodiment we bring to bear on the creation of metaphors does not mechanically and automatically lead to the emergence of universal or conceptual metaphors' (Kövecses, 2005, p. 287). Metaphor researcher Ning Yu (2008) argues in line with this that there is a dual connection of metaphor to both body and culture. Yu shows how an analysis of the distinction between primary and complex metaphors allows us to determine which aspects of metaphor are bodily or culturally based. This allows Yu to set up a hypothesis that primary metaphors, derived from bodily experience, are likely to be widespread and universal, while complex metaphors, based on conceptual mappings and cultural beliefs, are likely to be more culturally specific. The cognitive linguist Kövecses develops this argument in terms of a contrast between 'the universal' and the 'culture-specific':

> Some conceptual metaphors may be universal because the bodily experiences on which they are based are universal. Many of the same conceptual metaphors may reflect certain culture-specific features at a more specific level of conceptualization. Other conceptual metaphors may be entirely based on unique cultural phenomena (Kövecses, 2005, p. 177).

This would arguably be relevant not only for comparisons between cultures but also be of interest to study over time. For example, if our 'reality' changes—that is, the context in which we spend our lives, socialise and work—is it then likely that the meanings of some of our key concepts will also change, expand, or shift as well, particularly those that are 'culture-specific'? Will we then, to some extent, understand reality differently from before? This would mean that embodiment plays some part in a meaning-making process where metaphors will be used and meanings fluctuate as reality fluctuates.

The culture-specific meaning of concepts is likely also a potential source for confusion and even possible conflict. The same lingual expressions may, in fact, mean different things in different cultural contexts, especially in different languages. As Greiffenstern (2010, p. 4) points out, the Internet includes an English bias, leading to many non-English natives adapting to an online set of English concepts and phrases. Take for example the English word 'browse' and the expanded (and embodied) use of the concept in terms of 'browsers' with which you access the Internet. The 'browser' may very likely give a number of extended associations to the English-speaking user that is accustomed to meanings also relating to aspects other than accessing the Internet, namely inspecting something or reading something superficially by selecting passages at a random. For the non-fluent English speaker in language-areas that have imported the concept, such as Sweden and Denmark, the connotations of the concept 'browser', in the meaning of a tool for accessing the Internet, are likely much more narrow if used in its (still) anglicised form. It simply just means that tool for accessing the Internet. For the actions of reading something superficially, these languages already have existing concepts such as 'gennemse' (Danish) and 'bläddra' (Swedish). This means that the same lingual expressions can have (at least slightly) different connotations in different cultures (for an account on anglicisms angled toward Danish and Swedish, see Gottlieb, 2005).

Another example is the concept of 'forest'. In a comparative study on the meaning of the concept 'forest' in Sweden and in the eastern Canadian province New Brunswick, Andersson showed how the conceptions of 'the forest' and the role that it ought to have were different in the two areas (1994). For example, in Sweden the distinction between *natural* and *cultural forest* has become more prominent over the last years. In Canada, the rural values of forest were much less stressed than in Sweden, Andersson argues, which partly explained why the conflicts between the forest industry and the environmental groups were, in fact, much stronger in Canada and the opposition 'between cultural and natural forest' more emphasised (Andersson, 1994, p. 86). This means that 'forest' does not share any exact meaning, but is to some extent depending on both biological as well as cultural differences in, for example, Sweden and Canada's New Brunswick.

SEEING THE EMBODIMENT: JUSTICE *UNDER* LAW

Legal scholars such as Steven L. Winter, inspired by Lakoff and Mark Johnson, have spent significant effort in displaying the need for embodiment for law. That is, they claim that law needs to be talked and thought of as an object for us to be able to speak and think about it. For law and legal reasoning, this is a relatively new and contemporary development when related to this specific development in cognitive theory. Mark Johnson commented upon Winter's *A Clearing in the Forest* that Winter 'explores the way recent empirical research from the sciences of mind gives us a new understanding of legal reasoning as embodied, situated and imaginative' (Johnson, 2002, p. 951). According to Johnson, a key issue in relation to the embodiment of law becomes, then:

> How embodied organisms like us, interacting continually with our physical, social, and cultural environments, come up with laws and legal institutions that are at once constrained by our embodiment and at the same time are imaginative, creative, and flexible in their application to our ever-changing experience. (Johnson, 2002, p. 951)

This means that the conundrum or challenge is to understand how such a formally fixed entity as law, in its lingual expressions and principal content, relates in its meaning-making to a fluctuating context consisting both of social relations as well as human bodies and physical infrastructure in their widest sense. Winter, for one, argues that law is an abstract social phenomenon, and in its state of being an abstract, the need for metaphors is immense. Winter claims that law can constantly be found in a dependent position, which requires object-making, or reification, in which we talk about law as something physical (Winter, 2001, 2008).

For example, when we conceptualise understanding in terms of seeing, which is commonplace, this also follows a series of other closely linked expressions or associations that have to do with the condition *to see* (Lakoff and Johnson, 1980, 1999; Larsson, 2014b). For example, light, brightness, transparency, and clarity. Arguments can be *obscure* and difficult to understand; the *Enlightenment* followed a more scientifically grounded view of life; what long-term consequences an event can have can be *hard to see,* and sometimes we need to *look* at a problem in a different *light*. We can speak of *illumination* in theological terms, as someone reaching spiritual wisdom. In line with this, it is commonplace to speak of 'observing the law', which can mean to be orderly, in control, act correctly, etcetera. When it comes to EU law and its relation to the member states, a common conceptualisation is that the national law should be 'seen in the light of EU law', as in *Welfare Integration through EU Law: The Overall Picture in the Light of the Lisbon Treaty* (Damjanović and De Witte, 2008). So, we have one orientational metaphor, or at least the notion of a spatial relation ('under'), as well as the conceptual metaphor 'understanding is seeing' relating to light and transparency that helps us conceptualise law and justice. Let us continue with investigating a few more examples over time (see the conduit metaphors in chapter 2).

As mentioned, a way to see an indication of the changes over time in the use of particular phrases is to use Google Books Ngram Viewer[3]. Google has indexed much of the printed books and made the texts searchable (for over 5.2 million books until

3. For more on the details of Ngram Viewer, see http://books.google.com/ngrams/info (Last visited 24 August 2016).

2008). The Ngram Viewer is based on books written between 1500 and 2008 in American English, British English, French, German, Spanish, Russian, Hebrew, or Chinese. When you enter phrases into the Google Books Ngram Viewer, it displays a graph showing how those phrases have occurred in a corpus of books—e.g. 'British English', 'English Fiction', 'French'—over the selected years. As an adjustment for more books having been published during some years, the data is normalised, as a relative level, by the number of books published in each year.

The benefits of this method lies in the capabilities that the Google book-scanning project offers in terms of quantitative analysis of published texts (cf. Mayer-Schönberger and Cukier, 2013, pp. 83–6). While the potential is great, the method has a number of weaknesses too. The Ngram view only tracks words and phrases, which by no means directly reveal the underlying conceptual mappings. There is also the issue of representativity—for what may the published writings be said to represent? Nevertheless, the existence and popularity of a particular phrase or wording likely indicates its existence and popularity on a wider scale than merely regarding the selection that the Google book-scanning project offers and may thereby be indicating on how an underlying conception is expressed over time. At the very least, this will arguably indicate how views of law is expressed linguistically differently over time in the English speaking parts of the world, which will aid the conceptual analysis that has to be deduced from this source data.

Law as verticality

When the Swedish Constitution states that '[p]ublic power is exercised under the laws' (Regeringsformen 1974:152, section 1), it cannot do this without a spatial reference, ('under'), which demands some kind of contextual understanding from its interpretation. When scanning expressions of law in a wider field, both historically as well as geographically over different jurisdictions, it becomes evident that law often is seen as something that is *uplifted*, and that we, the people, the public, or even justice itself, is 'under' it. This notion or conceptualisation of law as something vertically above justice, may be seen as a specific case of the conceptual metaphor CONTROL IS UP, and one that likely has had a long presence in human culture. One example of the expression 'equal justice under law' can be found as a phrase engraved on the front of the United States Supreme Court building in Washington, DC. This phrase was first written in 1932 by the architectural firm that designed the building, according to one commentator to paraphrase an earlier expression coined by Chief Justice Melville Fuller (Peccarelli, 2000). The embodiment of 'control' as something that is above has its clear spatial roots in how human body orientation and its entailments are mapped onto more abstract experiences of control (Winter, 2001, p. 73). If we look at the indexed English books in Google scholar between 1800 and 2000, we see that 'justice under law' is a not very common phrase that albeit has increased somewhat during the twentieth century, with a peak in the 1960s, see the grey dotted line in Figure 3.1.

What does it mean if we were to understand it in terms of another specific relationship? When targeting the 'under' law, one can notice the meaning of the opposite, of being 'above' the law, just as Al Capone for a while seemed to be 'above the law'. 'Above the law' seems to have a steady usage, with a somewhat small peak in the 1850s, but a continuous application, see the solid black line in Figure 3.1.[4] I return to the 'legal boundaries' and 'within the law' under the section about 'law as area' below.

Law as object or body

The bodily representation of abstract legal thought is likely very old. For example, when Roman law was assembled in what became the influential *Corpus Iuris Civilis*, issued from 529 to 534 by order of Justinian I, Eastern Roman Emperor, it displayed the conception of law *as a body*, (cf. Murrow and Murrow, 2012, p. 276; Winter, 2001, p. 335; for a more specific account on 'law as a person' see Winter, 333f.). Further, 'body of law' is a common concept today, too. It seems as though the conceptualisation of 'law as a body' is long-term, albeit with the application of the specific terminology of 'body of law' steadily increasing from the second half of the nineteenth century with a corresponding decline of 'corpus juris' during the same time (see the solid and the dashed black lines in Figure 3.2). Interestingly enough, there was some kind of outburst in popularity during the 1870s regarding both of these terms, of which I have no explanation for (the 'outburst' is seen in both American as well as British English).

I have already discussed the 'object' of law, that is, how law can be conceptualised as an object in order to be thought of, taught, and talked about. There are several applications for this, of which 'body' can be seen as one, and the fact that it can be 'broken' is another one. I would like to discuss yet an additional example addressing specific laws or agreements that are of particular

4. An even stronger claim, although not referring to a spatial relation, would be the one Judge Dread, the fictional 'enforcer' in a dystopian future played by Sylvester Stallone, does when arresting an entire block: 'I am the law', *Judge Dredd*, 1995.

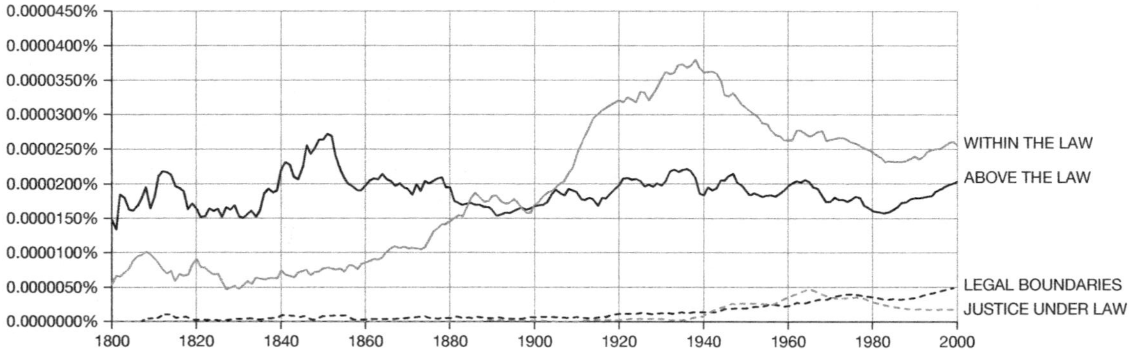

Figure 3.1 Google Ngram View of 'justice under law', 'above the law', 'legal boundaries', and 'within the law' from 1800 to 2000.

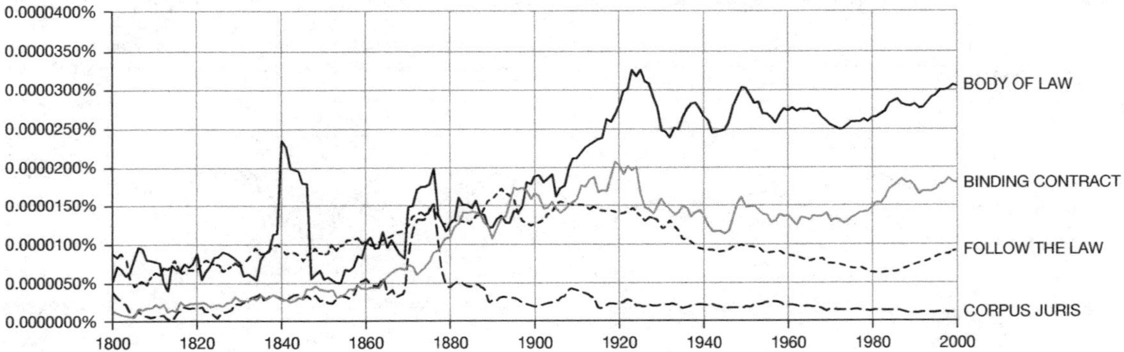

Figure 3.2 Google Ngram View of 'follow the law', 'body of law', 'corpus juris', and 'binding contract' from 1800 to 2000.

strength and hard to break or 'get out of': those that are *'iron-clad.'* On 26 December 1902, *The Florence Times* (original name of the Alabama newspaper *Times Daily*) reported that the 'City Council enact an iron-clad law' on the subject of the prohibition. The expression of 'iron-clad law', perhaps more common as an 'iron-clad agreement' or 'oath', with the meaning of a strong and definitive commitment, was a debated issue during the American Civil War. It was an oath promoted by Radical Republicans and opposed by President Abraham Lincoln to limit the political activity of ex-Confederate soldiers and supporters (Russ, 1934). This can be compared to the *'binding' contract*, as yet another embodied legal expression that, notably, has nothing to do with actual ropes or other physical measures of attaching someone to an agreement of some sort. The understanding of 'binding contracts' seems to have grown in the English-speaking world during the nineteenth century to be established and steadily used thereafter, see the solid grey line in Figure 3.2.

When it comes to expressions like 'following a rule' or 'following the law', it means some version of embodiment that refers to a spatial relation but also 'law as an object' that exists in this spatial 'room'. In the Ngram View, we see that 'follow the law' is steadily used in literature from 1800 to 2000, see the dotted grey line in Figure 3.2 (see also Winter, 2001, pp. 206–16). Although not so common in overall English literature, legal thinkers throughout the years somewhat regularly refer to law in terms of a 'path'. For example, the classic text *The Path of the Law*, by Oliver Wendell Holmes (Burton, 2000), is often referred to when debating the inertia of legal reform (Gillette, 1998) or the 'path dependence' of legal development (Larsson, 2011c). The practice of using paths as metaphors for something else is common, for example, leading to that they can be either 'straight' or 'narrow' (or long and winding), and metaphorically projecting the constraints of the source domain onto the target domain, in this case the law.

Law as area

Finally, there seem to be a commonplace understanding of moral or normative content in terms of an area delineated in a way that you can 'cross the line'. Law is here expressed as some kind of geographical entity with a spatial dispersion. That something is 'within the law' indicates that it is allowed, authorised, and lawful. The phrase 'within the law' is also commonly used from 1800 to 2000 with a steady increase, particularly during the four first decades in the nineteenth century. It is uncertain to what extent the expression was more commonplace in public culture, but it was, at least, during this time that the 1923 silent film drama with the same name was released (directed by Frank Lloyd and starring Norma Talmadge). The expression can be compared to the Swedish expression of being 'within the bounds of law' [*inom lagens råmärken*]. Additionally, if we look at the indexed English books between 1800 and 2000 in Figure 3.1 above, we see that the use of 'legal boundaries' is increasing, particularly during the late twentieth century, to something that looks like a potential future increase. Are we increasingly conceptualising law as something imposing boundaries?

DISCUSSION: LAW INCARNATE

Each of the expressions of law and justice mentioned in this chapter could be studied further and more thoroughly, and the use of the Ngram Viewer would need to be corroborated by other sources for the proof to be stronger. Together, however, they show the changing nature of how law is expressed and described, but not necessarily conceptualised, over time. It is clear that law as an abstract is highly metaphor-dependent. It has been stated that the conceptualisation of law as a body across time and (legal) cultures is strong and common, signifying a metaphorical and human *body* that incorporates into one body a variety of law—codes, statutes, and case law (Murrow and Murrow, 2012). An interesting part of Murrow and Murrow's account is the claimed link between this conceptualisation and the fact that it is not 'solely' a metaphor, that it 'emerges from and reveals the nature of non-conscious mechanisms of empathy in the brain that create a neurally simulated sense of sharing the feelings of others and thereby of being one-in-body with or virtually "equal" to them' (Murrow and Murrow, 2012, p. 276). Also Lakoff and Johnson's notion of embodiment is that the neurological system is wired to form meaning-making patterns from emergent spatiotemporal and kinaesthetic experience. Winter, for one, has in detail presented the theoretical background in this claim (Winter, 2001, cf. chapter 2). The notion of equal justice under law is a central part of the rule of law ideal that has been part of the Anglo-American system and others for centuries. Likely, a description of societal role of law is found in the conceptual metaphor of CONTROL IS UP and, hence, LAW IS CONTROL. The Greek expression of 'under law' (*hypo nomon*) is for example used by Paul in Galatians 4:21 and Romans 6:15. However, in the words of Kövecses, it is 'simplistic to suggest that universal aspects of the body necessarily lead to universal conceptualisation' (Kövecses, 2005, p. 294). In this case it means that even if there is no way around embodied conceptualisation, which likely has led to vertical conceptualisations such as CONTROL IS UP, it cannot be safely assumed that law in all cultures is understood as an instrument of control.

This means that the neurological connection to how we embody reality does not refute the fact that there is a clear cultural component to the conceptualisation of law—which my study clearly indicates in the sense that it seems to be a fluctuation over time in its use—and also suggests that there may be something more to be added to the analysis. This is also indicated by Lakoff and Johnson in terms of codependence amongst the body, context, and the sensorimotor system, with the result that it also contributes to shape how we reason (Lakoff and Johnson, 1999, pp. 42–4). The 'biosemiotic' perspective of Murrow and Murrow implies that there is a neural explanation to be found for why this particular metaphor is intertwined with our conceptualisation of law—that it has something to do with how the brain co-opts the body as a 'sign' for self, other, society, and law (Murrow and Murrow, 2012, pp. 294–302). By flipping this suggestion, one could perhaps interpret these results as having a link to the concept that how we conceptualise law can to some extent reveal how we understand the society we are part of. It may to some extent help us to understand how we conceptualise the relationship between the individual and the society, by studying the expressions of the conception of 'law as a body'. This is an interesting topic of discussion that of course would require much further attention to be fully investigated.

Even if we are neurologically wired to embody abstract concepts, such as CONTROL IS UP, this embodiment will to higher or lower extent be determined in each and every of its specific use by the culture in which it is used. So, even if CONTROL IS UP is a (near) universal embodiment shared by all earthly cultures, the fact that this vertical conception tied to control is connected to 'justice' as being subordinate to 'law' is a cultural construct.

Or, differently put, the conceptions of law and justice may be different in different cultures, and so the concepts themselves may be translatable or not, the relation between them may be similar or not, between cultures. It seems as if the relationship is common and widely disseminated over place and time, but that is not necessarily the same thing as that these conceptions and their complex relationship is universal. Again, remember the duality stressed by Kövecses:

> The human body, including its physiological, structural, motor, perceptual, and so on, makeup, is essentially universal (which is not to say that interpretations of the body and its workings or even many actual physical activities of the body are universal, as anthropologists have taught us). (Kövecses, 2005, p. 285)

In line with the fact that metaphors tend to highlight some specific aspects and hide others, one can argue for a raised awareness of what it means to *triangulate* metaphors in order to reveal conceptual bonds. For example, speaking of the body of law is in a sense an embodied way of speaking of the legal abstraction. However, the conventionalism of a particular use may hinder us from seeing the projection from one domain of meaning to another. We tend to become unaware of many or even most metaphorical concepts, which means that we tend to be unaware of the steering capabilities of the particular metaphors, the political side of their ontological claims, so to speak (Larsson, 2013a, pp. 133–34). By testing the conceptualisation with uncommon metaphors one may unlock this conventionalism and create awareness of the inherent meaning-steering aspects of a particular metaphor. This type of 'metaphor triangulation' may unlock hidden values or parts of the concepts that we commonly are not aware of when using the established metaphors. A sociolegal researcher that is quite known for his creative use of new metaphors to discuss and analyse law is Boaventura de Sousa Santos. He has used the Nietzschean three-step metamorphosis of the human spirit to display how law has developed in the modern era, from 'the spirit', to 'the camel' and finally 'the child' (De Sousa Santos, 1987, p. 279). What interests de Sousa Santos is the process of legal change and what these steps mean for legal development. He is even more clearly debating the inherent conceptions of law in *Three Metaphors for a New Conception of Law: The Frontier, the Baroque, and the South* (De Sousa Santos, 1995) where he argues that we are in the midst of a paradigmatic transition where 'the paradigm of modernity is undergoing a deep and irreversible crisis' (p. 571).

In line with this, the method of testing conceptions by replacing the metaphors they are most commonly reproduced by can be used in order to reveal how we understand particular phenomena, such as law or particular parts of law. Similarly the effect that this 'triangulation' brings is sometimes used in order to make a specific point, along with penning a punchy expression. For example, in a speech on 11 February 1851, the UK Prime Minister Benjamin Disraeli claimed justice to be 'truth in action', and thereby stressed that justice is not a passive quality. In that sense one can test the conceptualisation of abstract concepts, those that require embodiment. If justice often is conceptualised as being 'under' law, some light could be shed what this means for our understanding of the relationship between law and justice by conceptualising the relationship differently. What if justice were to be found 'next to' law? Or think of justice as being 'open to' law (justice is a container)?

Perhaps justice should be understood in the 'light of' law (seeing is understanding)?

CONCLUSIONS

From the conceptual metaphor-related perspective thus drawn, particularly bearing the embodiment in mind, it is *not* hard to think about law and justice as a physical as well as a social environment. In fact—the case is rather the opposite—it is probably impossible to think of such abstract concepts as law and justice without borrowing concepts based in a physical, spatial, and relational environment. In short, what we see is interpreted through cultural and social lenses to be used to bring order to an abstract 'inner' world. At the same time—which is an important point—in our instant interpretation and association, we do not sort out what is metaphorical from what is not. This means that literal meaning has no priority (Glucksberg, 2008). This is also the reason I argue for the benefits of 'metaphor triangulation'—that is, the replacement of prevailing metaphors with unusual ones in order to increase the level of awareness of what conceptual content the prevailing metaphors involve. How we describe law seems to take a number of forms that can be categorised based on how they are embodied from the physical and spatial. In this chapter the following three categories have been (abductively) concluded; Law as *a vertical relation*, law as *an object*, and law as *an area*. These categories link to the conceptual metaphors CONTROL IS UP and LAW IS A PATH, as outlined by Winter (2001). Over time, it seems, law is both seen as an entity of control and an impediment to motion: a general characterisation is that law is expected to exercise power (as in 'rule') in that it structures and holds accountable over time. Central for the conception of law, especially in relation to justice, is its vertical position in being uplifted, 'under' which justice—preferably—operates. It is not far-fetched to acknowledge the bodily representation of law—the *corpus juris*—as expressing the same inevitable need for concepts to express the abstract.

[4]

CONCEPTIONS OF COPYRIGHT

In this chapter I turn to the specific legal field of copyright and some of the challenges this field is facing in terms of 'piracy' or online file-sharing.[1] The question is: what type of metaphors do file-sharers employ to conceptualise copyright and its purpose in a digital context? Of interest is to study how they conceive the file-sharing community and how this 'online piracy' connects or does not connect to law, social norms, copyright enforcement, and computational traceability. Given the historical variations in the inherent emphasis on ownership and attribution in copyright law between an American and French continental context, are there, for example, noticeable differences between the American and the French respondents?

By drawing from CMT, this chapter analyses findings from a large-scale survey (20,000 respondents in this sample) on online file-sharing. The results indicate that copyright is not seen as 'property' by the respondents at all, that a majority of the US and French file-sharers would prefer to be more anonymous online in order to avoid legal enforcement, and that almost one out of five already uses such tools. The results indicate that there is a difference in how the American and the French file-sharers understand or conceptualise the future of file-sharing and its relationship to copyright: the French file-sharers focus more on

1. An earlier version of this chapter was published in *Lexis—E-journal in English Lexicology* in 2014, as "Conceptions of Copyright in a Digital Context. A Comparison between French and American File-Sharers" (Larsson, 2014a).

Conceptions in the Code. Stefan Larsson.

the actual artists, while the American file-sharers focus more on the role of the industry and the government.

METAPHORS WE FILE-SHARE BY

As developed in previous chapters, Lakoff and Johnson (1980, 1999) points out the absolute importance of metaphors for conceptual processes and abstract thinking. Theoretically and empirically, they counter a notion of the metaphor as simply an ornament of words bearing no deeper meaning for our thinking or our minds and show that the metaphor has a fundamental role in how our thinking and meaning making is conducted, stating that abstract concepts are largely metaphorical. This means that unlocking the metaphors constantly present in our language and mind—which in this chapter is argued to also be relevant to copyright law and norms in a digital context—may reveal to us how they are connected, what values and associations they bring, and on what conceptions they are founded. Arguably digitisation in society requires a remarkable need for metaphors and embodiment so that we can speak and think of the partly new phenomena that it brings with it (cf. Larsson, 2013c).

Although Lakoff and Johnson did not predict the massive transformations that digitisation brings, e.g. in what Castells described as a 'rise of the Network society' (Castells, 1996), the conceptual developments and challenges that digitisation presents are huge (Larsson, 2012a). Elsewhere I have studied (Larsson, 2013b) the conceptual development of the 'copy' as regulated by copyright, and will return to that study in chapter 5. The metamorphosis is immense, yet somewhat hidden, where 'the concept of "copy" has expanded to embrace a number of new phenomena consisting of a multitude of file formats for storing and distributing information in digital form' (Larsson, 2013b, p. 617). Thereby I demonstrate that the law is currently undergoing some sort of conceptual change without it being explicitly revised. Herein lies, perhaps, a key to understanding why copyright has such weak representation in social norms in a digital society, as discussed and studied by many (Andersson Schwarz, 2013; Feldman and Nadler, 2006; Karaganis et al., 2012; Lessig, 2008; Svensson and Larsson, 2012). When familiar words in a tangible context are similarly used to include actions in a digital environment, not only does this challenge our understanding of computer-mediated behaviour but also the laws that seek to regulate us—laws that may have been conceived in predigital circumstances.

In May 2012, The Swedish *Cybernorms research group* at Lund University in Sweden conducted a survey of over 96,000 respondents regarding file-sharing in collaboration with the infamous BitTorrent tracker The Pirate Bay. While this unorthodox collaboration is of interest in itself, I will here focus the empirical data resulting from it. As mentioned, it is the responses from the almost 20,000 French and US respondents that are analysed in this chapter. One section of the survey included an open-ended question that concerned future perspectives on the Internet and on file-sharing. This means that this data is a rich source for studies on how this near global file-sharing community conceptualises these issues and how they understand and describe both the current as well as the future situation via words and metaphors such as 'theft', 'sharing', 'market', 'surveillance', and 'law' (cf. Andersson Schwarz and Larsson, 2014; Larsson, 2014b; Larsson et al., 2012b).

The chapter's purpose

Given the challenges that intellectual property (IP) regulation has faced in a digital context, the purpose of this chapter is to present a study on how we may better understand how file-sharers conceptualise both copyright and the unauthorised file-sharing often called 'online piracy'. In order to achieve this aim, the chapter can be broken down into (at least) four central questions that simultaneously set some of the methodological and theoretical conditions:

Q1: What metaphors do online file-sharers use to conceptualise copyright in a digital society?
Q2: How do they understand property and intellectual property in this context?
Q3: How do they view the file-sharing community and how does this 'online piracy' connect or not connect to law, social norms, copyright enforcement, and computational traceability?
Q4: Given the historical variations in the inherent emphasis on ownership and attribution in copyright law in an American vis-à-vis a French continental context, are there, for example, noticeable differences between the American and the French respondents?

The first question concerns the explicit metaphors employed in this discourse. In this study, most of the material comes from the open-ended question in the large online survey. The second question corresponds to an ongoing debate concerning the extent to which IP regulation attempts to colonise or tap into how tangible property is conceptualised and regulated, and focuses on the P in IP in order to achieve more benefits for the rights-holders (cf. Herman, 2008). The third question is an attempt to study how the community regards itself, which is linked, for example, to conceptions of 'sharing is caring', technological determinism, and other modes of justification (cf. Andersson Schwarz and Larsson, 2014). Finally, the fourth question, which may come across as rather ambitious, addresses the fact that copyright and IP regulation have developed in cultures and traditions that differ somewhat, particularly in what is often regarded as the difference between the more economically bound and trade-related (sometimes described as utilitarian) tradition and the author-focused *droit d'auteur* tradition.

Metaphors and conceptions

In approaching an analysis of metaphors in copyright, the research on metaphors in cognitive linguistics constitutes our guideline. Of particular relevance here is Winter's *A Clearing in the Forest* (2001; cf. Johnson, 2007; Larsson, 2013a). Furthermore, this being the main point in Lakoff's and Johnson's work, we inevitably depend on metaphors for conceptual processes, which in this chapter will be further related to the notion of metaphor *clusters*, sometimes used in studies regarding this particular research object (cf. Larsson, 2011b, pp. 60–1, 72–3; Loughlan, 2006). Lakoff and Johnson focus on the conceptual metaphors that may have a number of *lingual expressions* which communicate the conceptual and underlying metaphor they relate to.

Copyright as property

One particularly durable strand of critique concerning copyright regulation addresses the romanticised notions of the individualist creator, often named the 'solitary genius' (Arewa, 2006; Larsson, 2012c; Patry, 2009; Rose, 1993). This individualistic view of authorship and the emphasis on the role of property rights for literary production is at the core of this legacy and has been key in explaining the purpose of, and justification for, the existence of a functioning copyright law (Fredriksson, 2009; Jakobsson, 2012; Rose, 1993; Woodmansee, 1994). Herman (2008) has analysed how copyright is increasingly described in rhetorical terms that indicate tangible property—i.e. conceptualised as property—which leads to stronger rights for the rights-holders.

This can also be seen in the metaphors that gain meaning in the debate when rights-holders, predominantly from the United States, speak of 'theft', 'trespassing', and 'piracy' when describing copyright infringing file-sharing (Herman, 2008; Larsson, 2011b; Larsson and Hydén, 2010; Loughlan, 2006). I've described this elsewhere as a 'colonising trend in copyright' (Larsson, 2011c) and is particularly relevant here in relation to the opposing conceptualisation of shared media content that the file-sharing community likely entertains. This development, which has mainly been linked to the American context, can already be noted in the way MPAA president Jack Valenti spoke of 'creative property' before the U.S. Congress in 1982 (Lessig, 2004, pp. 116f.). This is an example of how the pro-copyright industry constructs property-based claims on the intangible goods being copied in the digital sphere because it benefits their industry (Larsson, 2013b). According to media and communication researcher Peter Jakobsson, for example, the increased use of the concept of intellectual property can be understood through the development of increasingly neoliberal-influenced media politics (2012, p. 72), and Jakobsson argues that this is particularly true of the United States where the protection of private property has strong cultural roots.

Copyright infringement and social norms

Of interest here is the link between how phenomena are conceptualised and the norms that develop in relation to these phenomena. Within sociology and sociolegal research, emphasising the correlation between perception and action is not a new idea. For example, sociologist of law Banakar brings up Petrazycki's analysis of *intuitive* and official law and concludes that there is undoubtedly a link between perception (attitude, opinions, and beliefs) and action (behaviour and conduct) (Banakar, 2012, p. 15). This has also, from a more contemporary perspective, shaped theories on norms (social, legal, etc.) and their relation to cognition (cf. Hydén and Svensson, 2008; Svensson, 2008). This concept of norms has been used to measure the strength (or weakness, in fact) of social norms that correspond to copyright in a digital context (Svensson and Larsson, 2012; Svensson et al., forthcoming). Within the scope of this chapter, this is mainly of interest, as far as the conceptual aspect is concerned, to the extent that norms relate to conceptualisation such as conceptual metaphors and categorisation (cf. Larsson, 2013a). The link would then be that the way in which the digital environment mediates the sharing of movies, music, games, etc., is relevant in that the law is not perceived as normatively 'correct'.

France and the United States: Legal, lingual, and conceptual differences

In *Cosmopolitan Copyright* (2011), IP analyst and professor of library and information science Eva Hemmungs Wirtén studies the more than a century old historical background to contemporary copyright. She addresses the lingual differences in the time of the drafting of the 1886 Berne Convention for the Protection of Literary and Artistic Works, claiming that the Convention is a 'negotiation between legal systems, between copyright and *droit d'auteur*, between civil law and common-law traditions' (2011,

p. 11). With this, one might contend, she focuses on conceptual differences that accompany lingual cultures, and she further states that:

> One could argue that there have been two Empires in the history of international intellectual property relations: the French and the American. Both 'imperialisms of the universal' have cemented their power with the aid of an interface that is not so much territorial as it is symbolic: language. (Hemmungs Wirtén, 2011, p. 74)

Hemmungs Wirtén addresses Silke von Lewinski's (2008) *International Copyright Law and Policy*, in which von Lewinski concludes that 'it may be worthwhile studying whether the prevalence of the English language has had an impact on the perception of this field of law, or given rise to a possibly enhanced influence of "copyright thinking"' (2008, p. vii). This can be taken as support for surveying the conceptualisation of IP in French-speaking parts of the world in comparison with a US perspective.

For example, Jakobsson (2012) suggests that the different, geographically bound origins of liberal and romantic notions of copyright have had implications for copyright development in various countries (2012, p. 66). One can, therefore, distinguish between UK-US copyright in the utilitarian tradition and continental European *droit d'auteur* tradition (Samuelson, 1999). The utilitarian tradition attaches great importance to the benefits of copyright for the whole of society. Countries that instead follow the *droit d'auteur* tradition tend to emphasise the author's individuality and what is perceived as the special bond between the author and the author's works (Jakobsson, 2012, p. 66). For example, copyright in this tradition, as in France, often includes the less easily traded or sold 'moral rights' next to 'economic rights'.

In order to relate the possible lingual and conceptual differences between cultures and languages, the Swedish social psychologist and translator Johan Asplund (1991) can be useful. Asplund—who was inspired by Foucault and, to some extent, Lakoff—uses the concept of 'thought figures' to explain the difference between *translatability* and *understanding*. Asplund explains that when discourses have developed in different languages but are based on similar thought figures, or thought figures that both discourses utilise, the opportunities to understand each other are good, even if the exact translations of the actual concepts and terms are not possible. In contrast, if they develop their discourses in relation to thought figures that the other party does not utilise, they cannot (easily) understand each other. Asplund concludes that this could even be the case where two speakers of the same language are concerned (Asplund, 1991, p. 16). This may be compared with historical linguist Richard Trim's comparative theory of languages, which is more clearly influenced by Lakoff and Johnson and resembles Asplund's presentation. Trim claims that there are probably three main combinations of two basic forms: 1) two languages that share the same linguistic form and the same conceptual metaphor (this is then both translatable and understandable, to use Asplund's terms); 2) two languages that share the same conceptual metaphor but not the same linguistic form (it is not translatable but understandable); and 3) two languages that share neither, i.e. one conceptual metaphor may exist in one language with no equivalent in another, or they may have two different conceptual metaphors that convey the same figurative meaning (neither translatable nor understandable) (Trim,

2007, pp. 28–9). In short, Asplund and Trim seem to aim for a similar purpose, but from different scholarly backgrounds. It is argued, therefore, that Asplund's thought figures, to a considerable degree, can be linked to the conceptual metaphor theories of Lakoff and Johnson. It is not surprising to see that Asplund found inspiration in Lakoff's *Women, Fire, and Dangerous Things* (1987) (Asplund, 1991, pp. 10–12). However, as with every generalist claim that aims to categorise, there is always the empirical account that requires testing of the claim in particular cases. Asplund analysed a complex pair of concepts in Ferdinand Tönnies's *Gemeinschaft* and *Gesellschaft*, and it is of course possible that the more generalist claims accounted for above work better for complex concepts than simpler ones.

METHOD

In May 2012, three persons from the Cybernorms research group—Måns Svensson, myself, and Marcin de Kaminski—conducted an online survey that was accessed via a link on the front page of The Pirate Bay website. This survey was a follow-up to a similar survey we had conducted a year earlier. The study is sometimes referred to as the *Research Bay Study* due to the fact that during its 72 hours of operation, it replaced the famous Pirate Bay ship logo with another that depicted a magnifying glass over the ship (Andersson Schwarz and Larsson, 2014; Larsson et al., 2012b, 2014a, 2014b; Lindgren, 2016; Svensson et al., 2015). This chapter uses the data retrieved from the second survey (of three). During the 72 hours the survey was conducted, it received over 96,000 responses from across the world. More specifically, it is the responses from the almost 20,000 French and US respondents that are analysed here. The survey included an open section where the respondents could freely answer the following question:

> Please give us your own comments on the topic of file-sharing, especially how the situation in your home country looks and what you think will be the next big thing when it comes to the Internet and/or file-sharing.

It is the answers to this question that are used in the analysis, in combination with the opportunity to categorise the respondents according to downloading frequency, uploading frequency, geography, age, and gender. A comparison is made between the answers from the US and the French respondents. Given this specific site's character of being one of the most well-known facilitators for file-sharing through the BitTorrent protocol, it is likely that the respondents accessing the survey from this environment represent a file-sharing community. This means that this specific sample of file-sharers can tell us about a wider file-sharing community. What further corroborates that both the French and the American respondents are of a similar category can be seen in the fact that both groups share similar demographics in the study, e.g. in terms of gender, with a strong majority of males (French: 93.4%, American: 92.4%), and youth, where almost 80% of both the French and the US respondents are younger than 30 years old (almost half are aged 18 to 24).

There were 18,483 American respondents to the survey, and if you exclude those that never file-share, the number is 17,244. If you select those that download the most frequently, namely 'every or almost every day', the number of respondents is 4,730, which represents 26.8% of the US respondents (see Table 4.2). The total amount of respondents from France is 1,444, and if you exclude those that never file-share, the number is 1,271. If you select those that download the most frequently ('every or almost every day'), the number of respondents is 478, which represents 35.5% of the French respondents (see Table 4.2). I have chosen to focus on those who download the most frequently under the assumption that they are also the most likely to have embraced the digital environment the most strongly, may have the most elaborate justifications, and most likely do not share the norms that state that copyright infringement in online sharing is morally or socially wrong. In short, their actions reveal these respondents as being the furthest away from compliance to the requirements of copyright law.

Methodologically, the difference in size between the two samples poses a challenge. The American number of respondents being more than twelve times as high as the number of French respondents required on the one hand a normalisation of the data sets from the open-ended question so that they could be comparable, and on the other hand a testing of to what extent the differences in findings between the two groups were statistically significant. Regarding the normalisation of data from the open-ended question, the answers from the most frequent French downloaders consisted of 179,948 characters. In order to be able to compare to the most frequent US downloaders, a random selection was made among this latter group to match the same amount of characters (see Table 4.1 for further detail). A visual thesaurus called VocabGrabber was used to analyze the corpuses.[2]

2. Regarding the upload and download frequencies as well as the question of online anonymity, I have performed two-sample t-tests using a 95 per cent confidence interval to corroborate to what extent the differences are statistically significant. See tables 4.2, 4.3, and 4.4 for further detail.

Table 4.1 OCCURRENCES OF RELEVANT CONCEPTS DIVIDED INTO FRENCH AND US RESPONDENTS.[3]

Occurrences	*French resp.*	*US resp.*
1	Share (371)	Share (390)
2	Sharing (306)	File (341)
3	File (271)	Sharing (334)
4	Law (212)	Internet (239)
5	Internet (203)	People (183)
6	People (197)	Government (163)
7	Download (153)	Medium (138)
8	Good (116)	Law (100)
9	Money (99)	Free (95)
10	Free (96)	Industry (89)

FINDINGS AND ANALYSIS

The analysis in this section is combined with the presentation of data because they are so closely aligned. The subtitles represent the operationalised purpose presented above, that is, the four research questions.

3. The responses from the most frequent French downloaders consisted of 179,948 characters, while the number of responses from the most frequent American donwloaders was far larger. For this reason, a selection was made in the latter group's responses to a corpus of text consisting of the same amount of characters. This normalisation of the data was made in order to be able to compare the occurences of specific concepts in the responses from the two groups.

File-sharers' metaphors for copyright

When it comes to the metaphors the online file-sharers use that might expose how they conceptualise copyright in a digital society, the top list of occurrences reveals little since words such as 'have', 'but', 'not', etc. are of less analytic relevance here. However, a top list of the concepts that are of clear relevance to copyright and file-sharing displays more interesting traits (see Table 4.1). Both groups responded in English, and therefore, the words have not been translated.

Here we see some slight discrepancies that are of clear interest. For example, the fact that 'industry' and 'government' receive such a high rank in the US context may reflect the US debate overall in which the Hollywood industry and the American trade association representing the six big Hollywood studios and MPAA are so clearly present (see Table 4.1). Although industrial involvement is not insignificant in the French context, the link between the government and the industry is central in the US context. Table 4.1 reveals that the respondents tend to use metaphors for 'sharing' rather than any references to 'property' when it comes to understanding copyright. This also indicates that they are more inclined to speak, and think, in terms of 'copyright' rather than 'intellectual property', which is relevant not the least to the rhetorical colonisation mentioned above.

(Intellectual) property

When it comes to how the frequent downloaders understand property and intellectual property, one can begin by stating that the terms 'intellectual property' and 'property right' occur four times among the US respondents. They tend not to discuss this rights-based side of copyright in that sense. In the answers from the French respondents, 'property' occurs eight times and 'intellectual property' six times. Most of the answers are highly critical, and the French group often refer to the HADOPI law in very critical terms.[4] From this lack of conceptualisation of file-sharing as relating to property, we conclude that it completely opposes the US development mentioned above, where the media industry pushes for the 'property' aspects of copyright—the P in IP (Herman, 2008; Jakobsson, 2012; Larsson, 2014c).

Community and norms

First of all, this is a survey of the file-sharing community, of which more than half (56.7%) of the American respondents and almost 70% of the French respondents download more than once a week (see Table 4.2). This means that this particular group is highly unlikely to share the norm that copyright should be upheld in a digital context, at least not in terms of copy protection, which is different from, say, attribution (cf. Svensson et al., forthcoming; Svensson and Larsson, 2012). As mentioned, to test that the observed differences between French and US respondents are statistically significant, I performed two-sample t-tests. Using a 95% confidence interval, results suggest that all differences regarding downloading behaviour are statistically significant at the 10% level (see Table 4.2). Similarly, with the exception of respondents answering 'I do not know', the differences between

4. A French graduated response law was introduced in 2009 to target copyright infringers. It allowed for suspension of Internet access to a repeat infringer (but this particular part was revoked on 8 July 2013 by the French Government because that penalty was considered to be disproportionate). 'HADOPI' is an abbreviation of Haute Autorité pour la Diffusion des Œuvres et la Protection des droits d'auteur sur Internet.

Table 4.2 DOWNLOAD FREQUENCY BY THE US AND FRANCE RESPONDENTS

Download frequency	*US (% of this column)*	*France (% of this column)*	*Statistically significant difference*
Never	1,239 (7%)	77 (5.7%)	*
More than once a month	6,387 (36.2%)	330 (24.5%)	*
More than once a week	5,274 (29.9%)	463 (34.3%)	*
Every day or almost every day	4,730 (26.8%)	478 (35.5%)	*
Total	17,630	1,348	

the two groups are significant for all response alternatives when asking about the use of VPN services (see Table 4.4). However, test statistics indicate that the noted difference in uploading behaviour between French and US respondents only is significant when it comes to high-frequency users, i.e. those that upload material more than once a week or every day (see Table 4.3).

How frequently VPN or other, similar tools—i.e. technological tools that leave fewer traces in an online environment—are used (cf. Larsson and Svensson, 2010; Larsson et al., 2012b) may tell us something of the attitude toward legal enforcement in this context. This relates to 'computational traceability' in Q3 above.

We can see that both groups are similarly inclined to be less traceable online, although the French file-sharers to a higher degree use pay services for it. Interestingly enough, a majority in both groups would prefer to be more anonymous online (see Table 4.4), which could be interpreted as a clear indication that they do not share the norms that the law prescribes in this field.

Table 4.3 UPLOAD FREQUENCY BY THE US AND FRANCE RESPONDENTS

Upload frequency	*US (% of this column)*	*France (% of this column)*	*Statistically significant difference*
Never	11,433 (65.6%)	853 (64.6%)	
More than once a month	4,277 (24.5%)	303 (22.9%)	
More than once a week	1,020 (5.9%)	94 (7.1%)	*
Every day or almost every day	693 (4%)	71 (5.4%)	*
Total	17,423	1,321	

Table 4.4 USE OF VPN OR SIMILAR SERVICE TO PROTECT ANONYMITY

	US (% of this column)	*France (% of this column)*	*Statistically significant difference*
Yes, free service	14.4% (2,463)	10.2% (132)	*
Yes, paid service	5.9% (1,009)	8.8% (114)	*
No, but I'd like to be more anonymous online	59% (10,109)	56.4% (730)	*
No, I do not care about anonymity	12.4% (2,132)	16.9% (219)	*
I do not know	8.3% (1,419)	7.7% (99)	
Total	17,132	1,294	

US vs. France

For the US file-sharing community, it is far more common to speak of the industry and the government when thinking about the future of copyright. When conceptualising file-sharing and copyright, the findings among the US respondents indicate that they tend to think more in terms of how they relate to government and the entanglement of the industry in a societal context. This can be compared to the findings in the French file-sharing community, which focuses—in line with the historical attitudes outlined above—more on the artists.

Given the type of data in this study, however, it is hard to establish a more robust connection between national historical attitudes toward copyright and the file-sharing communities in the two nations. It is nevertheless still interesting to note that there are differences between the French and the American downloaders, which may be worthy of contemplation in, for example, further studies. In this regard, there are some concepts that stand out in comparison. In the answers from the French respondents, there were 78 occurrences of the word 'artist', 35 occurrences of 'industry', and 62 of 'government.' In the US respondents' answers, there were only 32 occurrences of the word 'artist', 89 of the word 'industry', and 163 of 'government'. This can be seen in quotes from the French file-sharers such as:

> 'I've never spent so much money in music, movies, and software since I've started doing voluntary donations to Creative Commons music/movies/series, to free and opensource software developers, and other artists I like'.

> 'I guess the biggest issue is supporting the artists/persons who released the file we share'.
>
> 'T.P.B. is one great place where I can download and listen freely to music and movies, and then buy them if they please me, because true artists deserve MY money'.

These are interesting and relevant concepts, and they reveal something of a difference between the two groups. This can be seen in quotes such as: 'The current trend is toward increased surveillance by government/industry' (American file-sharer). For others, there is a spirit of resistance and counter-activity in file-sharing or, at least, in their justifications (see also Andersson Schwarz and Larsson, 2014):

> Fuckers. The industry needs to embrace technology. If I had more legal options, I wouldn't steal. Sometimes a bit of disobedience is needed, I download the Metallica discography at least twice per week, then immediately delete it (Metallica isn't even that good), just to piss off Lars. (US file-sharer)

One of the earliest artists to famously oppose online file-sharing was the drummer in the rock band Metallica, Lars Ulrich (later nicknamed Lar$), who participated in the lawsuits against Napster in April 2000. This has become a symbol that still seem to live on in the file-sharing community.

It is more common in the American file-sharing group to speak of the 'market' (72 occurrences) and the idea that the industry should 'adapt' (69 occurrences), as compared to 42 occurrences for 'market' and 43 occurrences for 'adapt' among the French file-sharers.

CONCLUSIONS

The purpose of this study was to better understand how file-sharers conceptualise copyright and unauthorised file-sharing in a digital, online context. In order to do that, we used the open-ended answers from a large-scale online survey of the file-sharing community and divided the answers into one group of 17,244 US respondents and one group of 1,271 French respondents. The results indicate that there is a difference between how the French and the American file-sharers understand or conceptualise the future of file-sharing and its relationship to copyright. For example, the American file-sharers are more inclined to speak of the 'market', 'government', and the 'industry', and the French file-sharers are more inclined to speak of the 'artists' and their remuneration. This can, at least at a surface level, be seen in the terminological difference between 'intellectual property', the common concept in American copyright discourse, as opposed to the *droit d'auteur*—the author's right—in the French and continental copyright tradition. In other words, the findings can be interpreted to indicate that the US respondents are more inclined to conceptualise the issue in the context of a social dilemma that follows from a 'utilitarian' tradition, while the French norm and culture is in line with the notion that the creators should receive the benefits; this, then, constitutes the challenge the French respondents relate to, in a *droit d'auteur* tradition.

In addition, a majority of the file-sharers in both groups prefer to be more anonymous online which, in light of the fact that they know that file-sharing often is illegal, is a sign that the social norm that they are likely to act in accordance with is not in line with the legal norms of copyright. This may be connected to the fact that the file-sharers do not conceptualise copyright in terms of 'property' or similar notions, but rather in terms of 'sharing'.

[5]

COPIES

A Metaphoric Expansion of Copyright

Central for copyright—both in terms of its enforcement and how its role in the market and society is perceived—is the concept of the 'copy'. Along with how recording technology has developed, particularly emphasised in digitisation, the concept has not been a fixed one. It's meaning has changed over time. The chapter[1] utilises the conceptual metaphor theory outlined in earlier chapters to analyse how the concept of 'copy' in copyright law is expanding in a digital society to cover more phenomena than originally intended. For this purpose, the legally accepted model for valuing media files in the case against The Pirate Bay (TPB) is used in the analysis. When four men behind TPB were convicted in the District Court of Stockholm, Sweden, on 17 April 2009, to many it marked a victory over online piracy for the American and Swedish media corporations. The convicted men were jointly liable for the damages of roughly EUR 3.5 million. The case addressed how to calculate damages of file-sharing. For example, what is the value of a copy? The chapter uses a model for valuating files in monetary numbers, suggested by the American plaintiffs and sanctioned by the District Court in the case against

1. This chapter is a revised and developed version of the article 'Copy Me Happy—The Metaphoric Expansion of Copyright in a Digital Society', *International Journal for the Semiotics of Law* 26(3) (2013b), 615–634.

the founders of TPB, in order to calculate the total value of an entire BitTorrent site. These calculated hypothetical figures are huge—EUR 53 billion—and grow click by click which, on its face, questions some of the key assumptions in the copy-by-copy valuation that are sprung from analogue conceptions of reality and transferred into a digital context. This signals a (legal) conceptual expansion of the meaning of 'copy' in copyright that does not seem to fit with how the phenomenon is conceptualised by the younger generation of media consumers.

COPIES AND THEIR RIGHTS

When hearing Rob Reid's intriguing TED talk *The $8 Billion iPod*, it becomes inevitably clear that there is something ambiguous about the calculation of value of media products in a digital environment.[2] In the United States, Congress may designate a maximum $150,000 fine for 'wilfully' pirating a single copy of a single song under the Digital Theft Deterrence and Copyright Damages Improvement Act of 1999. This number is, according to Reid, 'grotesquely divorced from the actual damages and harm caused by a single instance of piracy'[3] and regarded as 'frequently arbitrary, inconsistent, unprincipled, and sometimes grossly excessive' by other critics (Samuelson and Wheatland, 2009, p. 441). The American music industry has sued more than 30,000 persons under this law, and since the consequences of

2. Rob Reid, TED talks: The $8 Billion iPod. See also the explanatory blog post Reid, The numbers behind the copyright math. 2012. [Blog post.] http://blog.ted.com/2012/03/20/the-numbers-behind-the-copyright-math/ [last visited 10 August 2016].

3. Rob Reid. (20 March 2012). The numbers behind the copyright math. [Blog post.]

losing would be devastating to the defendants, the majority settled without challenging the industry. As a result, the maximum $150,000 per song fine has never actually been imposed but is used as a power tool in the extremely asymmetric settlement negotiations. It does, in addition, address the notion of *copies* in copyright, and under what assumptions they should be valued and that this valuation be legally upheld in a digital society. In the Swedish case against the founders of the BitTorrent tracker The Pirate Bay (TPB), this question was drawn to its ultimate conclusion.[4] In District Court in April 2009, the four defendants were found guilty for assisting in violations of copyright law through the Pirate Bay site. They were sentenced to one-year prison terms and fined roughly EUR 3.5 million (SEK 30 million). Both sides appealed, and the three that were retried in the autumn of 2010 were all found guilty yet again and sentenced to higher damages of approximately EUR 5.4 million (SEK 46 million) though somewhat shorter prison sentences. In the TPB case, the figures per copy are exceptionally more modest than in the US regulation, and aim at what songs and movies cost in stores, as DVDs and CDs, but still rely on core assumptions that follow from an analogue and copy-based handling of media. These assumptions are tested in this chapter, via the calculation of the value of an entire *other* (anonymous) BitTorrent site, using a model sanctioned by the District Court of Stockholm in the case against TPB.

The copy-based conception of handling, controlling, and distribution of creative expressions is very homogenous in the global copyright regime (Larsson, 2011a, 2011c). In most jurisdictions, copyright owners have the exclusive right to exercise control over

4. Case B 13301-06, 17 April 2009, Stockholm District Court, and Case B 4041-09, 26 November 2010, Svea Court of Appeal.

copying and other exploitation of the works. The international treaties and directives focus the control over reproduction of the protected creation. For instance, the Berne Convention states that authors of literary and artistic works shall have 'the exclusive right of authorising the reproduction' (Article 9); the European Infosoc Directive speaks of 'the exclusive right to authorise or prohibit direct or indirect, temporary or permanent reproduction' (article 2); and TRIPS states 'the right to authorize or prohibit the direct or indirect reproduction' (Art. 14, section 2). The Rome convention provides us with an explicit definition of 'reproduction' in that it 'means the making of a copy or copies of a fixation' (Article 3 (e)). Further, Infosoc speaks of 'the rightholder of any copyright or any right related to copyright'. This motivates the analysis of 'the copy' as the central concept in copyright, tightly connected to a control over reproduction of copies.

Given that the conditions under which media are distributed have changed so fundamentally—from having been stored on various types of well-defined and delimited plastic objects to being disseminated as digital files in networks—the concept of 'copy' has expanded to embrace a number of new phenomena consisting of a multitude of file formats for storing and distributing information in digital form. These phenomena possess attributes that resemble those of the original ones, e.g. reproducibility, but also attributes that do not, e.g. non-materiality. The fact that copies in a copyright sense once only meant physical entities and now mean both physical as well as digital entities describes a conceptual development of 'the copy'. It is even possible to reflect upon a future where they no longer are perceived as physical entities. This may be more likely for some types of copies than others. For example, computer games may, in practice, be less of physical entities than books were at an earlier stage in time, but they are still regulated by the same legal construction. Much historical semantic change, as argued by Berkeley linguist Eve Sweetser, is attributable to metaphor: 'if a word once meant A and now means B, we can be fairly certain that speakers did not just wake up and switch meanings on June 14, 1066. Rather, there was a stage when the word meant both A and B' (Sweetser, 1990, p. 9). It is reasonable to think that for a number of concepts that are metaphorically expanded to fit in a digital existence, that stage is about now. However, we tend to need to reuse older concepts to describe new phenomena.

As outlined in chapter 2, Cass and Lauer (2004) focus on this conceptual transition in terms of CMT as a way to support the concept of *skeuomorphs*. They use the example of the transition from analogue photography to digital imagery and describe the conceptual borrowings and dependence of the analogue predecessor in order for us to make use of the digital variant. This comes with both advantages as well as disadvantages in terms of conceptual lock-in (Larsson, 2012b):

> When the new artifact is described using terminology from a prior iteration, this influences one's intent and encounter with the new artifact. The new artifact is initially understood using the norms and interpretive scheme of the old artifact, which aids in both transition to and adoption of the new medium. (Cass and Lauer, 2004, p. 252)

This means that CMT can be used to understand how the concept of copies is expanding to also include digital copies and in what sense this understanding of digital copies 'us[es] the norms and interpretive scheme' of the physical predecessors. To practically illustrate the problems associated with 'copy thinking' in a digital context, this chapter uses statistics from a file-sharing tracker site (similar to, but not, The Pirate Bay) and the monetary values per file that the Swedish District Court used in the case against The Pirate Bay (Case B 13301-06, 17 April 2009). The goal is to calculate what an *entire* site, such as The Pirate Bay, would be worth according to this established, legally sanctioned model. The sum would likely be enormous—demonstrating the ultimate logic of this method of valuing files. This is a way of criticising legal assumptions based on how linguistic and conceptual processes function. Even though the figures in the US example mentioned above may seem perverted in comparison to the more modest figures found in the TPB case, the latter still signals that there is something lost in translation (or impossible to translate) when the control-over-copy-conception is applied in a digital context. The purpose of the chapter is to better understand how copyright is challenged by digitalisation. Or, differently put, how current copyright in some aspects is ill fitted with contemporary digitalised reality. In this chapter I address this through five main questions:

Q1: How is the legal conception of a copy challenged in a digital environment?
Q2: To what extent is this relevant for nonlegal debates and rhetorical framing?
Q3: Which assumptions in copyright, based on analogue preconditions, are the most challenged in a digital context?
Q4: How is this relevant in relation to the difference between legal and social norms?
Q5: What can this case teach about change in people's conceptions of reality?

In order to answer these questions, this chapter argues, in line with the cognitive linguists that follow in the footsteps of Lakoff and Johnson regarding CMT, that our thinking is essentially controlled by metaphors and conceptions. A conceptual metaphor analysis of this legal construction holds similar benefits to the narrative analysis—namely that it 'uncovers hegemonic processes' (Halbert, 1997, p. 56). Images and conceptual frameworks largely determine how we understand our reality, and communicate and thereby fundamentally argue and debate. When speaking of framing—such a meta-metaphorical point—one must ask what types of thinking a particular frame *enables*. The frame will control what outcomes will seem reasonable (Lakoff, 2005). For example, in a review of Hu's *A Prehistory of the Cloud* (2015), journalist Hua Hsu is in the *The New Yorker* asking:

> What kind of thinking does the Cloud, so porous and diffuse, enable? Does our participation in the Cloud require us to surrender a bit of our privacy? Can it help explain the rise of the meme and our increasingly lax attitude toward notions of authorship and origins, the way something on the Internet begins to seem ubiquitous and ambient, as if it had always just been there? (Hsu, 2015)

The way that normalised and prevailing metaphors frame particular phenomena can also be viewed from the perspective of power. The person who maintains control of metaphors and conceptions will be able to shape the discussion to their own advantage. Having a background in sociology of law, I make the point that this dynamic also manifests in the formulation of law as a dialectical process that incorporates technological progress, social change, etc. Law also adds an aspect of formal power, in line with sanctioned metaphors and conceptions, which is why it is of particular relevance to understand just how law is functioning in relation to technological change.

THE PIRATE BAY CASE AND THE CALCULATION OF VALUE

Before going into details in the case of The Pirate Bay, I need to point out the rather obvious metaphorical content of the name itself. The point has been made (Larsson, 2011b) that the metaphor is double-sided in terms of being used both by the pro-file-sharing side to describe something rebellious and innovative as well as by the pro-copyright side to claim proprietary ownership over content. It is also a concept that is not completely fixed in its meaning. For example, the definition of piracy has, according to Litman (2001), changed over the relatively few years that it has been used to describe the activity of copying. Today, the metaphor is used to describe any activity that involves some kind of unauthorised copying, despite the fact that much of this copying is legal. 'Piracy' is by some seen as a transitional concept that may serve a purpose for a briefer time but is likely to have a diminishing role as the discourses around file-sharing calm down. I've elsewhere expressed that the term will likely be meaningful for the brief period of time when file-sharing represents something rebellious or otherwise deviant from a widespread and accepted value system (including one supported by law). When the flows of Internet become the defining paradigm, file-sharing is not likely to continue to be regarded as rebellious or deviant, and will therefore likely no longer fit well with the 'piracy' metaphor (Larsson, 2011b, p. 97).

The widely reported prosecution of The Pirate Bay led to a conviction in April 2009. The District Court found the four defendants guilty of complicity in violating the copyright act. The court referred to Articles 2 and 54 of the Swedish Copyright Act to the effect that anyone violating its provisions is required to compensate the copyright holder (Case B 13301-06, 17 April 2009, p. 85). If the violation was wilful, compensation was also to be paid for damages other than loss of payment. It is worth noting that damages were based on the loss that the person committing the *main violation* had occasioned the plaintiffs—rather than the complicity of the defendants. One crucial issue during the trial was, as mentioned, the value of a file. The trial primarily concerned 'complicity in violating the copyright act' in a specific number of cases—making 21 music albums, two songs, eight films, one television series, and four video games available to the general public (Case B 13301-06, 17 April 2009, pp. 57–58). In District Court, the four defendants were convicted and sentenced to one year in prison, which in the Court of Appeal became differentiated prison sentences for the convicted (see next chapter). In addition, they were to compensate the plaintiffs for their losses and court costs. Some of the loss involved 'reasonable compensation for utilisation' as prescribed by the copyright act. If the violation was intentional, 'compensation for other loss' should also be due (Case B 13301-06, 17 April 2009, pp. 99–101).

In the following sections I will first present how the court reasoned in relation to the calculation of value, and then proceed to, in my opinion, the most interesting model for such a calculation, put forward by a group of the plaintiffs. In order to understand the details of the inherent problems that this model leads to—a site can be valued at EUR 53 billion and grow click-by-click—I present how such a calculation is made according to this particular model approved by the court.

Finding a method for calculating these types of losses is no easy task and requires reflection and thorough assessment. The companies chose various methods of calculating compensation. In determining the size of compensation, a generally accepted legal practice is to use various rates (or the like) to construct a hypothetical licence fee. According to the Stockholm District Court, the absence of such rates does not mean that reasonable compensation is not to be paid. Under such circumstances, it is incumbent on the court to assess the matter (Case B 13301-06, 17 April 2009, p. 95). The Nordic film companies decided to use a hypothetical fee—and the court approved their model. The model was based on the assumption that the Swedish market had six million Internet users, and that neighbouring countries had approximately the same number. An expert with 'long experience as a senior executive in the Swedish film industry' stated that a reasonable licence fee for the three popular Wallander films would be approximately SEK 700,000 (approx. EUR 80,000) and that a reasonable licence fee for the less popular *Pusher 3* would be SEK 150,000 (approx. EUR 17,000) (Case B 13301-06, 17 April 2009, pp. 96–97). The geographical focus on Sweden in that model is not particularly elaborated upon. It is safe to say that TPB has had a very Swedish focus, not the least due to the fact that the founders are Swedish and much of the initial activity seems to have been located in Sweden. But since the site shifted to English, and given the global popularity of the site, a calculation model based on Swedish consumption is not really rooted in the factual situation. For example, in a survey on the file-sharing community that we in the Cybernorms research group conducted via the site, it was concluded that slightly over half of the 76,000 respondents were located in Europe (54.7%) and 27.7% were located in North America (Svensson et al., forthcoming; Larsson et al., 2012b). The Court, to a great degree, seems merely to be cutting the Gordian knot in accepting that calculation model.

Although this calculation model proceeds from a number of assumptions—and reveals an ingrained concept that a legal case must be linked to a single country—the second calculation model is arguably of greater interest from a metaphor perspective. The Swedish record companies and American film companies used a method of identifying what they 'should have been paid' for each individual download of their copyrighted works (Case B 13301-06, 17 April 2009, p. 98). Use of this calculation method makes several things obvious. The record companies stipulated a theoretical loss of EUR 6.50 per album and EUR 0.70 per song multiplied by the number of downloads (Case B 13301-06, 17 April 2009, p. 97). The American film companies proposed an approximate cost of between SEK 222.50 (approx. EUR 26) and SEK 261.47 (approx. EUR 30) for the films and SEK 415.81 (approx. EUR 48) for the television series. The District Court found that to be too high and lowered the prices to SEK 150 (EUR 17) and SEK 300 (approx. EUR 35), respectively (Case B 13301-06, 17 April 2009, p. 99).

Statistics from one of the other popular BitTorrent tracker sites in autumn 2009 can be used to assess the underlying assumptions of 'copy' in copyright. Approximately ten billion downloads had been performed via this other website (statistics that were shared with the author with an agreement that the name of the BitTorrent tracker should not be revealed). What the site's files would have been worth if one of the models had been used that the District Court deemed reasonable in the case against The Pirate Bay can thus be calculated. However, the court questioned the reliability of the counter on The Pirate Bay's website as evidence for establishing the number of downloads. This was only one of the three reasons cited for reducing the number of downloads used to calculate damages by 50%; the other two reasons had to do with geographical considerations and judicial analogies from illegal file-sharing to other rights of disposition under copyright law performed in connection with calculating the value of copies. Based on the conclusion that half of the number specified by the counter was probable, the District Court found—with certain misgivings—that the actual model for calculating the value of damages was reasonable, which—all things considered—validates the copy metaphor in this connection.

To avoid an obvious criticism of my use of download statistics from another BitTorrent site in calculating value, I will also reduce my figure by 50% to just under five billion downloads. I am doing so even though not all the reasons for which the District Court reduced the number of downloads by 50% are relevant here. My point—that enormous amounts can be calculated and in some ways constructed—will still be eminently clear. Assuming as the District Court did that a large majority, say 90%, of downloads entail copyright violations, a quick calculation shows that at least 4.42 billion illegal downloads were performed via this anonymous site.

$$(9{,}828{,}507{,}127\,/\,2) \times 0.9 = 4{,}422{,}828{,}207.15$$

If the statistics from the anonymous BitTorrent tracker are followed, it is obvious that most downloads were film, music, and television series. However, several assumptions have been made for the sake of clarity. In the first place, it is reasonable to stick to the assumption that 90% of downloads were illegal, regardless of category. Not because it by necessity must be true, but to follow the assumptions sanctioned by the District Court. A total of 27.99% of illegal downloads were films and 25.01% were music. Thus there were 1.24 billion illegal downloads of film and 1.11 billion illegal downloads of music on the anonymous BitTorrent tracker. TV series accounted for 20.67% of the offering, which translates to 914 million illegal downloads. The results presented below are rounded off to the nearest integer.

Film: 4,422,828,207.15 × 0.2799 = 1,237,949,615
Music: 4,422,828,207.15 × 0.2501 = 1,106,149,335
Television series: 4,422,828,207.15 × 0.2067 = 914,198,590

If these results are then used to calculate 'reasonable compensation for utilisation' in the manner applied by the District Case in the case against The Pirate Bay according to the claims of the record and film companies, the amounts start to be rather high. If compensation is based on SEK 150 multiplied by the number of downloaded films, the result is approximately SEK 186 billion (approx. EUR 22 billion). If a corresponding calculation is made of music with the additional assumption that 75% are albums, for which the District Court approved a price of EUR 6.5, and 25% are individual songs, for which the District Court approved a price of EUR 0.7, the current exchange rate yields a value of SEK 52 billion (approx. EUR 6 billion) for albums and SEK 1.9 billion (approx. EUR 120 million) for individual songs—or close to SEK 54 billion (EUR 6.5 billion) for music as a whole.

Film: 1,237,949,615 × 150 = 185,692,442,277
Albums: 1,106,149,335 × 0.75 × 62.7 = 52,016,672,460
Individual songs: 1,106,149,335 × 0.25 × 6.75 = 1,866,627,002

In addition, TV series can be valued at SEK 300 each (EUR 35). My somewhat rough approximations require an assumption, since I do not know what percentage of downloads were individual episodes, an entire season, or a number of seasons. But a reasonable assumption is that a majority of people download entire seasons—so that a minority download single episodes. The assumption here then is that 80% of downloads are entire seasons and 20% are single episodes. Entire seasons may seem to be overbalanced by this approach, but this is counterbalanced by the omission of single episodes in the calculation below, particularly because the District Court did not determine their price. The District Court valued the individual seasons at SEK 300—which yields a total of approximately SEK 219 billion (approx. EUR 25 billion) based on my calculation.

$$914{,}198{,}590 \times 0.8 \times 300 = 219{,}407{,}661{,}700$$

To sum up, the total value of illegally downloaded music, films, and television series from a popular BitTorrent tracker site similar to The Pirate Pay can be calculated at approximately SEK 459 billion (approx. EUR 53 billion) in accordance with a legally sanctioned model based on the reproduction of copies.

$$\begin{aligned}&(\text{film})\ 185{,}692{,}442{,}277n + (\text{albums})\ 52{,}016{,}672{,}460n\\&\quad + (\text{individual songs})\ 1{,}866{,}627{,}002\\&\quad + (\text{television series})\ 219{,}407{,}661{,}700n = 458{,}983{,}403{,}440\end{aligned}$$

EUR 53 billion is a lot of money. It is more than half the budget of the Swedish public sector in 2014. But it is nevertheless the value that the calculation model sanctioned by the District Court—based on the copy metaphor under copyright law—yields for a single BitTorrent tracker (in 2009). Not only that, but I have reduced the number of downloads by 50% for the sake of my mathematical exercise, which is only partially relevant in comparison to the District Court's reasons for reducing The Pirate Bay's statistics by 50%.[5] If the argument for 'other financial loss' awarded to

5. Considering that my goal has been to show the kinds of amounts that may be arrived at by using the price per copy approach in connection with file-sharing statistics—i.e., not calculating on the basis of Swedish file-sharing per se or whether the judicial analogy from illegal file-sharing to other right of disposition under copyright law actually works—it would have made just as much sense to skip the 50% reduction that the District Court applied to the number of downloads in the case against The Pirate Bay. Another significant difference is that the plaintiffs in the case against The Pirate Bay specified a figure based on a simple counter, whereas I have allowed the site owners themselves to provide the statistics.

the plaintiffs in the case against the founders of The Pirate Bay were also applied and the proportion of the loss in terms of other damages were to form the basis of the calculation, the damages would increase by approximately 50% to some SEK 688 billion (approx. EUR 80 billion). The above figures are huge however you slice them, but are nevertheless a logical consequence of the District Court's calculation method. How should these results be interpreted? Depending on the perspective, the focus can be on copyright law, i.e. faith in the construction that revolves around the value of individual copies. That would reasonably cause dismay about the enormous amounts of money that the copyright industry is losing out on. The logical, but somewhat hasty, response in that case would be to demand even stronger measures against file-sharing in order to recover such enormous income. The alternative interpretation is to consider whether there is a fundamental flaw inherent to translating the concept of the value of copies to a digital context.

ANALYSIS: THE PROBLEM OF REGULATING COPIES

The relative reasonableness of the damages awarded in the case against The Pirate Bay stems from the choices that the plaintiffs made. However, it is obvious that those who sued the men responsible for the site intentionally focused on only a small number of works—and a relatively limited period during which file-sharing occurred. Let us not forget that these numbers carry with them a *semiotic content*, symbolising something as they are presented (which is evident not the least in Rob Reid's intriguing TED talk mentioned in the introduction of this chapter). If many works and more copyright holders had been assembled into a campaign, the amounts would have been so large that an outcry by the general public would certainly have ensued. The Swedish tradition for awarding more modest (in relation to, for example, the United States) damages would clearly have been set aside. Whether that would have influenced the actual judgement of the court is another matter—legally speaking, the dichotomy of *right* and *wrong* is just as valid regardless of the amount—but it would likely have compromised the perceived legitimacy of the judgement and shed light on the dilemma posed by attempting to value files in specific monetary terms.

Q1 The law and digital copies

If we begin with the central theme in this book, cognition researchers have demonstrated that metaphors are necessary to thought processes. Metaphors provide us with images that allow us to talk about abstractions; they are deeply rooted in our language and mental outlook. Conceptual metaphors are providing structure to our associative thinking. Nevertheless, these structures are not value-neutral: conceptions frame thought structures that bring out functional aspects of some phenomena while concealing or diverting others. A tricky part is found in the lack of awareness of the power of metaphors and conceptions, the fact that they tend to be rather modest and retiring, that they are taken for granted in our language and remain out of reach of our everyday awareness—while still being in part a construct which may have a number of alternatives that could distribute power in various ways. Depending on how copyright is conceptualised, debates and arguments follow naturally in its wake (Andersson Schwarz and Larsson, 2014; Herman, 2008; Larsson 2011c, 2012a; Loughlan, 2006; Patry, 2009; Yar, 2008). Potential legislative amendments will always be hopelessly trapped within the logical walls inherent to prevailing ideological structures. Once the leading but underlying conception of copyright has morphed from a balance of mutual interests between creators and the general public to a system that focuses on the rights of copyright accumulators, the response to an essentially dysfunctional copyright system is to introduce more sanctions, more control, stricter punishment, and further criminalisation of associated behaviours (or technologies) (Larsson, 2011c). Critics end up in a position where—trapped in the prevailing notion on which regulation is based—they have to explain why they are against artists, musicians, and authors. No matter that this at least partly is due to a way of framing the issue (see Q2).

It is in this sense that the concept of copy in copyright is renegotiated in a digital context to also include digital copies conceptualised in a similar way as for physical copies. This negotiation is rhetorically underbuilt by other concepts mainly relating to physical property in order to make this renegotiation seem just and the following means for enforcement seem meaningful, which I develop more below (Q2). However, I've attempted to use CMT in order to clarify the conceptual development that is at hand, and to support the argument that digital and analogue copies are phenomena based on essentially different conditions, but still sharing a few similarities that keeps this conceptual bridge (legally) intact. The above mentioned media development researchers Cass and Lauer use the term 'skeuomorph' (see also chapter 2) in order to analyse how conceptual development is occurring in a metaphoric fashion (see also Larsson 2011b, pp. 62–6, 101–04). The skeuomorph is metaphoric in the sense that the design or aesthetics of one phenomenon (such as analogue photography) is reused when talking and thinking about a similar but still different phenomenon (such as digital imagery):

> When new artifacts are presented to the public, many times they are described with metaphoric allusions that are grounded in prior iterations of that artifact. These metaphors assist people in their transition to understand and use new technological processes and artifacts. (Cass and Lauer, 2004, p. 252)

This means that it is in many ways very meaningful to speak of and conceptualise digital copies in a similar fashion as we do with analogue copies. This is however in some particular aspects also *not* meaningful, and can be described as a conceptual lock-in that prevents us from clearly seeing where the conceptual expansion is failing. To the extent that the value of a copy should represent costs for manufacturing it, that is to some extent meaningful for physical products, but not meaningful in a digital context. There are constraints also to the expanding metaphor (or skeuomorph):

> After collective learning about the new artifact that occurs, artifact users may discover new functionalities beyond what the 'transitional functions' suggest and places where the metaphor breaks down (Cass and Lauer, 2004, p. 252).

There are other reasons for the legally enforced per-copy-valuation, and some of them are historical and derived from its analogue background. We must however be aware of the construction to see where it reaches its boundaries.

Q2 Rhetorical and conceptual consequences

The manner in which metaphors and conceptions act on a subconscious level harbours a restrictive, conservative, and difficult-to-criticise tendency, which makes this type of analysis important, not the least in order to 'uncover hegemonic processes', in the words of Halbert (1997, p. 56). Herman (2008) has analysed the metaphor of 'copyright is property' and shown how rights-based characteristics borrowed from something that is a well-established in an analogue, physical, and cultural sense (ownership—primarily property—rights) have also been brought to bear on copyright issues. One consequence of the rhetorical use of this metaphor is that hackers who circumvent technical obstacles to copying are described as having 'broken in' (Herman, 2008, p. 1). This context very much relates to common ideas of (nonintellectual) property. Herman's analysis signals a loan to the copyright debate of rights-based characteristics of analogue, physical, and culturally well-founded ownership, especially in real estate (see also Patry, 2009, chapter 6, and McLeod, 2007, who speaks of a 'simulation of property', p. 275). The consequence of the rhetorical use of this metaphor is that it becomes natural to talk about someone 'trespassing'—i.e. hacking technical barriers—and 'stealing' in the sense that they are copying, or sharing, computer files. Herman shows that the property metaphor dominates the general mental image of copyright, and therefore much of the debate and sometimes even the thinkers and scholars who seek to reconceptualise the challenge that digital content offers for the law. Metaphors can serve as persuasive tools that frame complex issues in a particular way, in this example resulting in a pedagogical and rhetorical advantage for those who propagate the conceptual links to the ownership of physical things (Yar, 2008). This, in turn, preserves the idea of copies, but also gives a similar rhetorical advantage to frame debates in terms of 'theft', 'pirates', 'parasites', 'trespassing', etc. (Larsson and Hydén, 2010). That is, they are actions based on analogue conditions and physical objects but are metaphorically transferred in order to define new types of actions under digital conditions (Larsson, 2011c, pp. 102–104, 2012c; see also Loughlan, 2006). Even if 'trespassing', 'piracy', and 'theft' are to be found outside copyright law in a strict lingual sense, they support the conception of copyright as tangible property, and frame the conception of the 'copy' as an entity that logically requires legal protection in terms of time, access, reproduction, and distribution (Andersson Schwarz and Larsson, 2014).

Q3 Challenged assumptions

Copyright is a global regulatory structure supported by international treaties, as well as EU and Swedish law. This chapter maintains that the ostensibly insignificant aspect of copyright law that focuses on monitoring the number of copies produced is a manifestation of an idea or conceptual metaphor that embraces much of the copyright system—and that is in conflict with current reality. The conclusion of the District Court in the case against the founders of The Pirate Bay that proper compensation for the plaintiffs was approximately SEK 30 million (approx. EUR 3.5 million)—a figure that at least is somewhat possible to comprehend—stemmed from a calculation based on the value of only a small number of works. Applying time-honoured analogue conceptions for calculating value to the categorisation of digital flows in terms of copies gave rise to a fundamentally different practice—which would still generate absurd, astronomical amounts if the reasoning were pursued to its logical conclusion.

As mentioned, the global dissemination of copyright has led to relative homogeneity in most of the world. This has been accomplished through international agreements (such as the Berne Convention and TRIPS Agreement) and harmonisation within the European Union (such as the INFOSOC and IPRED directives), as well as legal collaboration among the Nordic countries, etc. (Larsson, 2011b, 2011c). Some of this construction concerns control of accessibility and artistic works, but more importantly the control of copies granted to the copyright holder. According to Article 2 of the Swedish Act on Copyright in Literary and Artistic Works, which is relevant in the TPB case, 'Subject to the limitations prescribed hereinafter, copyright shall include the exclusive right to exploit the work by making copies of it and by making it available to the public . . . .' This provision can be viewed as an idea that was born out of, and worked well in, an analogue world where reproduction and distribution of copies related to actual physical artefacts. Under such conditions, legal protection for control of copies was clearly based on the manner by which the material was disseminated. The law kicked in when someone printed unauthorised copies, often with the intention of selling them, given that such activities were costly. The prevailing assumption seem to be that people listen to music and watch films, as well as consume and store media, according to the patterns that prevailed in an analogue world. For instance, nobody could conceivably purchase the same album twice, much less ten times. Or, as is probable in BitTorrent file-sharing, download the complete Beatles collection just to find a handful of songs you like and erase the rest (or the entire Metallica collection every week, only to delete it, just to 'piss off Lars'; see chapter 4). An assumption in the per-copy-calculation of value is that downloading something for free with a click is a sure indication that it would have been purchased under other circumstances—occasionally for quite a lot of money. Even though economic theory would state that the value of a work would have little to do with the cost of reproduction or distribution the argument is used by the pro-copyright side in the debate as something leaning toward a 'right to be paid' (Loughlan, 2006; Patry, 2009, p. 78–9). There is sometimes expressed a notion in the cultural industries that artists (voiced by the middle men, the production companies) *should* be able to live off what they create. Economic theory would in this regard be brutally honest and state that if nobody is willing to pay, then the value of the creations is zero.

This can be seen as that law is artificially supporting a partly very weak market. The market, however, may be weak only if one looks at the focused buying and selling of actual movies and music (etc.) but not necessarily if you view it as a consumption including broadband access, merchandise, live concerts, movie theatres, and a whole range of technological equipment such as smartphones, iPods, and computers, that is necessary for taking part of the cultural artefacts. This, of course, has been observed by the content industry the last few years, which has to do with a restructuring of the market in terms of ISPs selling stream-solutions, computer developers starting with music sales, etc.

The dilemma associated with controlling copies also embraces the well-established dichotomy between public and private. According to Article 12 of the Swedish Copyright Act, 'Anybody is entitled to make, for private purposes, one or a few copies of works that have been made public'. However, it is fairly obvious that the line between public and private has radically shifted in a digital context. A number of researchers have called attention to the focus of copyright on copies and the problems associated with specifying the right to control the number produced (Larsson, 2013b; Lessig, 2008). That was reasonable in an analogue context when every copy was a physical object but creates a slew of difficulties when translated to a digital context. Even commercial actors that perpetuate a flow-related metaphor such as streamed media must adopt the concept of copy when it comes to counting the number of times that a consumer has clicked to download files. Agreements with copyright holders rely on this approach. The idea of copy dominates within this framework.

Q4 Social norms and copyright

Whatever conclusions are drawn about this particular case of copies in copyright, the figures in the example above reveal that illegal copying of audio and video media (legal or otherwise) is in many ways socially acceptable behaviour in contemporary society, as concluded in a number of studies (cf. Altschuller and Benbunan-Fich, 2009; Feldman and Nadler, 2006; Larsson and Svensson, 2010; Larsson et al, 2012a; Svensson and Larsson, 2012). The question then arises as to the democratic value of disseminating such media. How can such a powerful social norm be handled when it does not correlate with the way that such behaviour is currently regulated?

Can legislation that promotes stronger and expanded copyright prove successful in practice? If so, at what price? A number of studies reveal that measures are taken by file sharers to be more untraceable online (Beekhuyzem et al., 2011; Karaganis et al., 2012; Larsson et al., 2012a) especially as a response to harsher copyright enforcement (Larsson and Svensson, 2010; Larsson et al., 2012b). Is it even possible to secure compliance with such regulation when social norms are so different? Furthermore, the political question arises as to whether such regulation should even be attempted if legitimacy and democracy are to be given due consideration. Whom is legislation designed to protect? The general public? Culture as an end in and of itself? Market structures and middlemen? A more general question about the legitimacy of the law can also be posed, if, for example, the law is risking the criminalisation of a major part of the young generation.

Q5 Change in conceptions

Litman's *Digital Copyright* (2001) argues that a kind of metaphorical progression—i.e. a shift in underlying assumptions—formed the basis of U.S. copyright legislation in the 20th century (2001, pp. 77–88). It began with a fairly narrow perception of the rights that authors and originators were to be granted, and increasingly focused on a reciprocity model that regulated the relationship of a creator to the general public. In the 1970s, copyright began to be regarded more as an incentive model. Litman maintains that the idea of negotiation has gradually been replaced by a model that proceeds from an economic analysis of legislation to define copyright as a system of stimulus measures. She asserts that the success of this model stems from its simplicity, given that it presumes a direct correlation between the scope of copyright and the number of authors whose works are produced and distributed. This framing or logic states that stronger copyright by necessity leads to more creativity, and is also confirmed by professor Halbert in *The State of Copyright*: 'The assumed starting point for politicians is that all forms of intellectual property are good and that the government's role is to increase regulation of these property rights until the threat posed by those who "steal" intellectual property is alleviated' (2014, p. 26). This particular regulatory approach, no matter that it can be questioned on a number of accounts, is tightly wedded with the control of reproduction and distribution of copies.

Born Digital by John Palfrey and Urs Gasser (2008) argues that the new generation has a completely different attitude to music than the generation that was introduced to the Internet at a later stage in life. They do not think of recorded music in the form of LPs, audiotapes, or CDs that can be bought at a record shop. Music for them is a digital phenomenon that can be downloaded, transmitted, and shared with family and friends. This is where the 'flow' metaphor acquires cogency, as a polemic against the idea of copy and its control. If we believe the narrative of those who have grown up in the embrace of this omnipresent digital network, it will shape our view of a number of contemporary phenomena related to IT policy, copyright, and file-sharing. In a recent study on file-sharing and media behaviour, which included focus group interviews with 15-to-16-year olds, the conception of how media is represented became evident (Svensson et al., forthcoming). All participants in the focus group study claimed that music and movies were an important part of their lives. Yet, no one claims to even remember having actually paid for digital purchase of either individual tracks or full albums. And furthermore, physical CDs represents a forgotten medium. One could therefore argue that parts of the conflicts emanating from the legal regulation of copyright today can be described in terms of a battle of conceptions. The analogically-based conceptions regarding the importance of the control over reproduction of copies battles with digitally based conceptions regarding flow of media where copies in themselves are not of the same importance. I've earlier addressed that it is the 'the conditions of reality that have changed quite drastically, when it comes to the possible reproduction and distribution of media' (Larsson, 2011b, p. 125), but that these new conditions can be conceptualised in different ways. And an important difference between law and society when it comes to conceptions has already been mentioned, and lies in the 'fixation' of conceptions—or at least lingual expressions linked to specific conceptual understandings of a given phenomenon—in or through law. I've elsewhere stated that conceptions are in a sense 'more "liquid" in a social context, which means that there can be a conflict in conceptions of reality between the legally embedded and the socially entrenched and distributed' (Larsson, 2011b, p. 125).

An enticing futuristic perspective emerges: the role of the Internet in human life, its emerging character, and the posture assumed by contemporary policy. What is happening now—and what will it be like later on? This forward-looking approach harbours an inevitable perspective that concerns generations not the least in the sense that members of the next generation have been born as users of information technology without necessarily regarding themselves as technologically competent. It is simply interlaced with their lives. Their existential notions—as well as their conceptual metaphors—are rooted in this context. Although a flow-based exchange does not necessarily eradicate the concept of 'copy', reflection outside the existing, legally formulated way of thinking becomes all that more urgent. Nor is it necessary to present one of the concepts as more correct than the other in some kind of objective sense. What is interesting is how such metaphorical thinking works in an empirical setting. Each of the conceptual models offers certain advantages and disadvantages. However, one of them is legally sanctioned and established as part of a powerful, worldwide regulatory process.

CONCLUSIONS: COPY ME UNHAPPY

In the transition from analogue to digital society and all the associated changes in social patterns, work organisation, public policy, democracy, creativity, and reproduction—not to mention the role of the law—mental images, metaphors, and their underlying conceptions will alternately be turned topsy-turvy and remain constant. In order to fathom the social norms and behavioural changes that are in the making, as well as the legal and political challenges involved, greater understanding is required of the ways that language and law interact. Some legal elements work well in the new context, some not so well, and previous thought patterns should not be allowed to hold sway to the point that new ones cannot be recognised. This means that metaphors are far from innocent. Those that the legal system employs are inevitably linked to power in one way or another. The law is a ubiquitous instrument of sanctioned governmental control. Successful legal regulation not only prescribes behaviour but includes conceptions about how the world is and should be arranged. However, a digitally distributed world calls into question centralised power as the legitimate interpreter of right and wrong. Social norms that have emerged in relation to behaviour based on the changes that the Internet has brought about do not have to be consistent with the way that analogue technology worked before the advent of the Internet.

Ian Brown, professor of Information Security and Privacy, concludes in his contribution to the 2015 version of *Sage Handbook of Intellectual Property* that 'large rights holders have spent nearly two decades trying to alter the nature of the Internet and the personal computer to fit business models relying on scarcity and the control of copies, with little success' (Brown, 2015, p. 581). The more recent development is however that legitimate market solutions see the light, even if complaints are heard that license negotiations take an inordinate amount of time.

In this case, copyright in a digital society describes a case of conceptual legal change of metaphorical importance. This means that the digital and physical copies that copyright regulates are in many ways similar, yet significantly different phenomena. The important difference lies in that they do not share the same constraints. The copyright-dependent (US) industries may make use of this conceptual confusion between intellectual property and property rhetorically. In order to speak and think of the abstract entities, they become embodied and conceptualised in physical terms. The legal conception of copies and the importance of control over their distribution and reproduction may be less meaningful to those, often younger, that are used to the sharing of copies in a digital environment. This fact may be part of the explanation for why the social norms correlating to copyright are weak in these groups. In brief terms, the essence of 'culture-bearing products' is conceptualised as more volatile or flow-like than that which is regulated in the control-over-copies-based copyright (remember how Gibbs describes conceptual metaphors as both cognitive and social, 'in the sense of facilitating people's social behaviours'; Gibbs, 2014, p. 30).

This means that there are legal and policy implications following from the flow-metaphor and the conceptualisation of the world it represents. First, it does not necessarily make sense to count copies. The number of copies is not by necessity a logical conception to regulate in a digital setting. Second, and related, the effort to control copies in contemporary IP is misdirected. The artificiality of the boundaries that are created in order to maintain these boundaries and protections are likely not seen as an outcome of natural constraints by large segments in society, and that tends to boost the perceived illegitimacy of the law. This is also true regarding geographically bound jurisdictions on such copies. In short, the law is not regulating reality as reality is perceived. The distance between the social conceptualisation and the legal conceptualisation is wide enough for the social contract to be challenged in this case. This probably supports the idea that if there is any willingness to pay for media other than this willingness, it would relate more to a flow-based conceptualisation of content consumption, which we also see in the growth of subscription-based streaming services. An analogy would be TV and radio.

If one were to speculate on the gap between social and legal norms, with regards to copyright from a cognitive perspective, it seems as though the aspects of control over copies that most contemporary regulation is based upon just does not correlate well with how the younger generations understand reality in terms of media consumption. This would then mean that in order to reinstate an existing version of contemporary copyright, without a change in its underlying conceptions, to be fully functional the law must not only appeal to rational decision making (and the fear of getting caught) by an extreme enforcement (Svensson and Larsson, 2012) but this must also be complemented with a whole set of constraints to be enforced upon the digital environment. This environment likely needs to be locked in and limited to function in a similar fashion to the material reproduction and distribution of copies so that people eventually conceptualise this environment as naturally limited and constrained.

[6]

PLATFORM, STORAGE OR BULLETIN BOARD? THE SWEDISH PIRATE BAY COURT CASE

The Swedish court case against the founders of The Pirate Bay, TPB, is of interest not only in terms of addressing underlying conceptions of 'copy' in copyright, but also in the sense that a key issue in the case regarded the branding of TPB, i.e., the analogical characterisation of what TPB 'really' is. The outcome of this branding would define to what extent the founders and caretakers of TPB were active or passive in relation to the factual sharing of copyrighted content that was catalysed by TPB. This chapter uses conceptual metaphor theory, CMT, to further develop the concept of *skeuomorphs* (reuse of old concepts for new phenomena) in order to analyse the Swedish Pirate Bay court case.[1] The chapter thereby shows how this theoretical foundation can be useful for the analysis of digital phenomena, that are largely understood through metaphors and skeuomorphs. Also, when attempting to understand and conceptualise new digital phenomena such as TPB, law in a digital society is inevitably affected. Hence new phenomena can be fought over in a 'battle of metaphors' in the TPB court case, e.g. the arguments of seeing TPB as 'a platform', 'bulletin board',

1. An earlier version of this chapter was published as 'Metaphors, Law and Digital Phenomena: The Swedish Pirate Bay Court Case' in *International Journal of Law and Information Technology* 21(4) (2013c), 354–379.

Conceptions in the Code. Stefan Larsson.

or an 'impure search engine'. This was of primary relevance for the outcome of the case.

CATEGORISATION, DIGITALISATION, AND LAW

On 1 February 2012, the Swedish Supreme Court denied the appeal for trial, which let stand as the final ruling the judgment in the lower courts against the founders of TPB, including prison sentences and grave damages. This case, as we've seen in earlier chapters, displays a number of significantly interesting questions that can be found in the intersection of social change, technological development (not least in terms of digitalisation), and, most importantly, our understanding of both digital phenomena and law. Central to this chapter, therefore, is understanding digitalisation in terms of conceptual change—how metaphors and *skeuomorphs*[2] are necessary to understand digital phenomena—as well as guiding our conceptualisation of the particular phenomenon. This also plays a role in legal processes—that is, in the practice of categorising phenomena in legal circumstances—and is here elaborated further than with regards to the specific concept of 'copies'. Therefore, *the purpose* of this chapter is to develop a deeper understanding of to what extent metaphors and skeuomorphs are relevant for law in a digital context. The case displays a number of the features relevant for understanding conceptual change and practices of categorisation relevant for such a context. For example, a key question to address in terms of conceptualising new phenomena concerns what TPB *is* (or *was*, for the time period that was

2. From Greek: *skeuos*—vessel or tool, *morphe*—shape.

of relevance for the trial). How was the legal process struggling with this question—analogously and metaphorically? Of particular relevance, here, is the use of concepts such as 'platform', 'search engine', 'bulletin board', and (the founders of) TPB as entrepreneurs, while at the same time describing an activity by relying on embodied metaphors such as 'torrents', 'domain names', 'links', 'swarm', and many more.

In order to address and answer these questions, I use cognitive theory on conceptual metaphors, outlined above. As mentioned, a key feature of metaphors is that one concept, from the source domain, is projected onto other concepts, the target domain, in order to create an effect of meaning. CMT, as outlined by Lakoff and Johnson, includes the notion of an experiential base of understanding that borrows concepts from bodily or spatial references to bring order to abstract phenomena—often referred to as embodiment. Of extra importance for this thesis is the legal analysis that has been made based on the work of Lakoff and Johnson (see chapter 2 above), particularly when it has been concerned with both law and technology (see for example Blavin and Cohen, 2002; Herman, 2008; Hunter, 2003; Larsson 2013b, 2013c, 2015).

A key notion underpinning the argument in this chapter is the need for studying the particular implications that the digital environment brings for law. This can be related to the 'law of the horse' debate from the second half of the 1990s in which Judge Frank H. Easterbrook[3] argued for the general perspective—the lack of need for a specific 'cyberlaw' field—and Lawrence Lessig was the most prominent counterpart (1999a, 1999b, 2006, see also Reed, 2012). Lessig pointed out a number of ontological issues— e.g. the essence and implications of

3. The origin is a 1996 cyberlaw conference presentation by Easterbrook, which he later published—see Easterbrook's *Cyberspace and the Law of the Horse* (1996).

the coded architecture—that he allowed to testify to the need for specific legal attention to not only what the digital domain brings to law but also, in fact, when it acts with legal normativity in itself: when code, de facto, is law. Although I argue for a specific interest in what digitalisation means for legal concepts, the conceptual legal change can by no means be narrowed down to be relevant only for digitalisation—it is just that this specific shift is very broad, happens over a short period of time, and is near-global, which makes it particularly interesting to study. From the perspective of the classical sociolegal scholar Karl Renner (1949; see also Robson, 1977; Larsson, 2014c), it is clear that the legal concepts or expressions may remain the same, while their function will transform as society transforms, no matter to what technology or reason the transformation can be linked. I return to this in terms of property in the following chapter.

When it comes to law, technology, and metaphor, this chapter does not specifically address the common-law legal reasoning, but will likely be of strong relevance nonetheless due to the extensive analogising from precedent that the common law tradition brings. When reusing relevant principles from the past and applying those principles to new settings, the conceptual bridges between concept and phenomena may be either stretched or reinterpreted, sometimes with a certain amount of creativity on the one hand or more strict conceptual path dependence on the other (Bellia et al., 2011, chapter 2). Patricia L. Bellia and others address these specific challenges regarding old concepts and metaphors when faced with a legal issue in a digital matter and stress that we must be 'acutely aware of the metaphors that lawyers and judges employ when thinking about Internet legal issues' (Bellia et al., 2011, p. 26). The early (in terms of the Internet) analyses of metaphors of relevance for our understanding of digital issues relate to the concepts that describe them in the first place, such as 'web', and 'cyberspace' (as a place) (cf. Blavin and Cohen, 2002).

Unlocking the metaphors constantly present in our language, minds and—as is argued here—law can reveal to us how they are connected, what values and associations they bring, and on what conceptions they are founded. In line with this, there has been, since the 1970s, a growing awareness of the importance of understanding the practices of *categorisation* within cognitive linguistics. Scholars like George Lakoff have, via empirical evidence, strongly emphasised that how we conceptualise the world is deeply dependent on categories, and that categories are—as opposed to a common and rationalistic view—not necessarily 'clearly defined, mutually exclusive and collectively exhaustive' (Lakoff and Johnson, 1999, pp. 373–414). In fact, they claim that categories are graded (they tend to be indistinct at their boundaries) and inconsistent in the status of their constituent members (cf. Amsterdam and Bruner, 2000; Winter, 2001). These practices of categorisation, I argue, become evident whenever a new phenomenon emerges, one that previously lacked a name. This is of further significance in a digital society, or rather, in a society in the process of becoming more digitalised and interconnected through massively embedded technologies such as the Internet (Larsson, 2012a). When a new phenomenon actually emerges—as with many digital phenomena—it will likely be judged and understood in terms of already present phenomena and their already accepted concepts. For example, Cass and Lauer (2004) describe the connection between analogue photography and digital imagery in terms of the latter as a *skeuomorph* of the first.[4]

4. The term *skeuomorph* is used in archaeology, for example, to describe the phenomenon of when pottery has imitated metal objects (Kingery, 1993), or when tombs have been prepared with home-like features in order to satisfy a culturally

Digital imagery bears enough resemblance to analogue photography to be accepted as a mere iteration of it, despite the fact that it only shares a few attributes and constraints of the former phenomenon.

THE PIRATE BAY

The Pirate Bay is interesting not the least due to that it is likely the most famous and resilient BitTorrent tracker in the world. It was started in 2003 and has constantly been increasing in numbers of visitors, regardless of any conviction against its founders or attempts to block it in national legal regimes, with a decrease only from 2014 or so. According to the site-ranking company Alexa, in May 2008, TPB entered the top one hundred of the most visited sites globally and by November the same year had twenty million unique peers for the first time since the site was launched.[5] In April 2011, it had climbed to number 86 and in May 2012, it had reached place 63 for the most visited sites in the world with more than five million registered users. Even if it suffered from going offline for brief periods late 2014 and early 2015, and has fallen in popularity.

embedded belief that the dead need these amenities on their 'journey' (Bloch-Smith 1992, p. 188). Furthermore, the skeuomorph concept has been used in design studies for similar borrowings of style or shapes that may lead associations from a new object to an older and already established one. I use the concept in order to analyse the (conceptual) development of copyright as a legislation, with its key concepts developed in pre-Internet times facing tremendous problems when the same concepts are used in a digital environment (2011b).

5. See Torrentfreak 18 May 2008, http://torrentfreak.com/the-pirate-bay-100-popular-080518/ [last visited 14 October 2016] and the web information company Alexa.

On the 17 of April 2009 four men were sentenced to one-year prison terms and fined roughly EUR 3,500,000, (SEK 30,000,000), for assisting in violations of copyright law through the Pirate Bay site. Both sides appealed, and the three that were retried in the autumn of 2010 were all found guilty yet again and sentenced to higher damages of approximately EUR 5,400,000, (SEK 46,000,000), but somewhat shorter prison sentences (26 November 2010, Case B 4041-09). The fourth defendant did not show up and was reportedly ill and living in Cambodia at the time. On 1 of February 2012, the Swedish Supreme Court denied the appeal, which meant that the decision in the Svea Court of Appeal stands. It is the proceedings in both the District Court as well as the Court of Appeal that are analysed in this chapter.

There is a growing amount of literature on, and analysis of, The Pirate Bay, from a number of perspectives. There are articles on technological aspects (Adrian, 2006), entrepreneurial and disruptive aspects (Elert et al., 2016; Palmås et al., 2014), the political implications (Miegel and Olsson, 2008) and from a legal perspective, on IP law in general (Touloumis, 2009; Kirchberger, 2011; Lewen, 2008), the case in particular (Bogdan, 2010; Carrier, 2010) and the aspect of secondary liability for intermediaries (Andreasson and Schollin, 2011; Helman, 2009). One contribution that is relevant to understand TPB is the Swedish anthology *Efter The Pirate Bay,* ('After The Pirate Bay'), from late 2010 (Andersson and Snickars, 2010). This anthology, edited by the media researcher Jonas Andersson Schwarz and the then head of research at the Swedish Royal Library Pelle Snickars, included many of the key Swedish writers on digital politics and Internet research. There are some articles in this anthology that deal with metaphorical aspects of TPB, whether in terms of a court case or as a political and technical phenomenon. For example, Nina Wormbs grasps the overall picture by analysing how technology itself bears a metaphorical function; Jonas Andersson Schwarz's contribution breaks down the Internet into two types of metaphors, the fishnet versus the spiderweb; Peter Jakobsson discusses 'the commons' as a widespread metaphor that has been used to criticise the organisational and legal structures of the media world; and Daniel Johansson (2010) analyses the flow metaphor 'music like water', which I elaborate on in the same anthology, to highlight the absurdity of an estimate of the value of *digital* copies based on a model originating from valuation of physical copies (see chapter 5).

Another contributor, Leif Dahlberg, has elsewhere elaborated on his contribution dealing with law and its metaphoricity, as he discusses the TPB verdicts in relation to 'liquid spatiality that characterizes the activities of the pirate' and the 'juridical liquid space that is created in his or her wake' (2011, p. 262). Rasmus Fleischer devotes part of his contribution to an overview of how different forms of water over the years have been used as metaphors for what happens in computer networks (2010, pp. 259–80). Fleischer continues with a characterisation of TPB, which is of relevance to this chapter: 'Question: What was the Pirate Bay? Answer: during the first decade's second half, The Pirate Bay was a wide variety of things—that somehow held together', and describes this in terms of 'assemblage' (Fleischer, 2010, p. 263).

Copyright law in a digital society has been extensively debated and criticised over the last fifteen years or so. In the late 1990s, a highly critical perspective toward what intellectual property regulation meant for creativity in a digital domain emerged among a group of American scholars. Law professor James Boyle (1996) was early in identifying copyright as one of the crucial issues in the construction of the information society in *Shamans, Software, and Spleens: Law and the Construction of the Information Society* (see also 1997, 2008). He was followed by scholars such as Lessig

(1999, 2004, 2008), often dealing with aspects of the coded architecture or the implications of how easily remixes are done in the new medium; Jessica Litman (2001); and Said Vaidhyanathan (2001), who all criticised different aspects of IP in relation to the digital. Copyright has been increasingly criticised from several perspectives, for example, from the perspective that the legal norms do not reflect social norms, which may pose a democratic issue to legislation (Reed, 2012, pp. 121–28; Feldman and Nadler, 2006; Karaganis et al., 2012; Svensson and Larsson, 2012). It is also in this light the analysis of the TPB case can be viewed, bearing in mind that there is something about digital developments that has led to a systemic challenge to IP regulation which, according to some commentators, is linked to the fact that we experience reality differently—that is, the matter that copyright used to regulate is not the same matter anymore. As stated, this chapter uses the case of TPB and its encounter with legal procedure in the case against the four men held as being behind TPB in order to show how important the metaphorical process is to law. Of exceptional relevance for this particular court case, here argued, is how a conceptual reuse in a translation of analogue concepts to describe digital phenomena is imperative for the outcome in the case.

One importance of metaphor research here lies in what I pointed out as the dangers of metaphor not being perceived as metaphorical. When the metaphors are not perceived as metaphors, the hidden conceptual mappings and framings will lead to what is perceived as the only possible alternative for the purpose of a given regulation. Any attempted revisionary arguments will then be framed within the prevailing conception, no matter what arguments are produced. Thi
the conception is analytically unlocked and displ
metaphors that reproduce it. This means that leg

as well as legislation, may be framed and conceptualised in a particular way without us even seeing alternative frames or conceptualisations. Some legal scholars, such as the above mentioned Linda Berger, have used CMT to study hidden values in decision-making in court. Berger's conclusion is that lawmakers cannot avoid being affected in their decision making by myths, metaphors, and symbols (2009, pp. 262–66). Berger studies cases that concern determining which parent is most suited to be granted custody when there often is no rational basis to choose one parent over the other. Decisions must, however, be taken and instead of doing this under the false impression that those decisions are pure and objective, Berger sees the solution in raising the awareness of how meaning is produced in the legal field. William Patry (2009) is yet another legal scholar who accentuates the importance of metaphors, in his case in what he describes in terms of the 'copyright wars'. He is very critical of how the American copyright lobby has managed to use metaphors surrounding copyright to their advantage. The combination of Berger and Patry can therefore work to highlight the importance of studying hidden values within the decision making in the court case against The Pirate Bay, which so clearly is linked to the 'copyright wars', and therefore includes a number of metaphors either for or against the protective elements in copyright.

THE PIRATE BAY AS A METAPHORIC COURT CASE

When the Svea Court of Appeals convicted the founders of TPB in November 2010, it was for aiding in crime against Swedish Copyright Law on the grounds that this service promoted the sharing of copyright material without the authors' consent (Case B 4041-09). The wording in chapter 23, paragraph 4 of the Swedish Penal Code reads:

> Responsibilities in this code are prescribed for a specific act, it shall be imposed not only on those who carried out the deed, but also the one that facilitated this by giving advice or assistance. The same shall apply in relation to what in another law or regulation is a criminal offense for which imprisonment is prescribed.[6]

The problems and interesting issues that the case against TPB displays do not necessarily concern internal legal relations (although, there are, in my opinion, interesting issues here as well) as much as the relationship between law on the one side and social and cognitive structures on the other. Only when law is *not* placed in a societal context may it be possible to perceive copyright law as internally coherent and therefore nonproblematical. The media researcher Andersson (Schwarz) offers a characterisation of TPB by its links to and entanglements with different actors and actions:

> A service like TPB is 'sticky'; the abundance of connections and interrelating actions acts like glue, and ties both venture capitalism, party politics, partisanship, and even the idea of stewardship of a sensible cultural ecology into its vortex. TPB was never simply one thing; it can be seen as a conservator of a mainstream cultural supply, as well as a radical opponent

6. Generally referred to as 'BrB 23:4': 'Ansvar som i denna balk är föreskrivet för viss gärning skall ådömas inte bara den som utfört gärningen utan även annan som främjat denna med råd eller dåd. Detsamma skall gälla beträffande i annan lag eller författning straffbelagd gärning, för vilken fängelse är föreskrivet'.

> to the big media corporations, as a harbinger of free media distribution or, conversely, as a hedonist absorption of mere self-gratification—and much more. (Andersson, 2011)

This act of interpreting the existence and identity of TPB is relevant also as a legal activity performed in court. And the outcome of this activity, this labelling of metaphors, has direct consequences for the outcome of the case, for the judgment of guilt or non-guilt, and for the measurement of penalty and damages. It simply, but directly, ties to the cognitive approach of labelling abstract concepts while dealing with what could be termed a 'conceptual battle' that took place in the TPB case. One central question in the Pirate Bay case was the question of what the Pirate Bay *is* (or *was* in the time frame selected for the case) and to what extent this existence could be regarded as either *actively* participating in copyright infringement or merely *passively* aiding the behaviour of the users of the site. Much of the liability was tied to a debate centring this dichotomy of active or passive participation, and the battle in the case was to a large extent fought through attaching different metaphors to TPB. This conceptual debate has its roots in argumentation that employs different metaphors to frame characteristics in an active or passive mode.

Sometimes, TPB is described as a *platform* where others carry out their sharing. For example, in the District Court case, one of the defendant's legal counsellors, Jonas Nilsson, pointed out that the copyrighted works did not pass through the Pirate Bay's computers: 'All the Pirate Bay does is passively provide an empty platform'.[7] This characterisation is also relevant for the services the Pirate Bay depends on in terms of Internet access, broadband,

7. Day 12 in first instance of case nr B 13301-06: 'Allt Pirate Bay gör är att passivt tillhandahålla en tom plattform'.

and storage. By labelling TPB as a 'platform', the defendant connects to a rather rich field in which digital intermediaries have strategically used this metaphor. 'Platform' is what Tarleton Gillespie (2010) calls a structural metaphor that 'though it may go unnoticed by the casual listener or even the speaker, gives the term discursive resonance'. In this particular extended case, a legal middle step is also developed when the aspect of time is added to the characterisation if the passive contributor of a service neglects to avoid participating in illegal actions, even though the contributor has become aware of their illegality. The middle step is found in the passive contributor becoming conscious of its contribution. All of which leads back to what the Pirate Bay *is*. For example, Monique Wadsted, of the law firm MAQS that represented a range of major movie labels and computer game companies in the TPB case, claimed in the Court of Appeal that TPB does not represent a 'passive site'. She claimed that the categories on the site make it easier to find torrents and, therefore, the site has made it easier for users.[8] On the other hand, one of the defendants, Fredrik Neij, claimed that 'The Pirate Bay is a transmission service, not a storage service'. Therefore, the defendants should not be liable to any copyright violations that occurred via the site due to the release from liability that service providers receive through the E-Commerce Directive.[9] He also claimed that TPB should be regarded mainly as a search engine.[10]

8. Day 2 in the Court of Appeal, 29 September 2010, see Svenska Dagbladet http://www.svd.se/naringsliv/digitalt/dag-2-svdse-rapporterade-direkt-fran-the-pirate-bay-rattegangen_5422679.svd [last visited 23 August 2016].

9. Directive 2000/31/EC of the European Parliament and of the Council of 8 June 2000 on certain legal aspects of information society services, in particular electronic commerce, in the Internal Market (Directive on electronic commerce).

10. Day 2 in the Court of Appeal, 29 September, 2010, see Svenska Dagbladet http://www.svd.se/naringsliv/digitalt/dag-2-svdse-rapporterade-direkt-fran-the-pirate-bay-rattegangen_5422679.svd [visited 23 August 2016].

So, should TPB be seen as a 'search engine' or a 'bulletin board'? To what extent should the founders be seen as 'entrepreneurs'? Is it meaningful to speak of TPB as an 'assemblage'? The latter is relevant not least in terms of the generative nature of the underlying technology. To label a set of functions in transition is to shoot at a moving target which, in this case, has legal consequences when valuing the meaning and significance of the case for the future.

TPB as a search engine

The question of seeing TPB as a search engine or 'search service' was debated in both the District Court as well as in the Court of Appeal. For example, in the District Court, the prosecutor asked the expert witness Kristoffer Schollin if TPB is a 'storage service'. Schollin replied, 'Yes, but only a very limited one' and preferred to call it a 'search engine', (or 'service' in direct translation), as the defendants had described it in interrogation.[11] Present in the case was the question of how to relate that to the fact that a technology in a sense is neutral, that it can be used for both good and bad (legal and illegal) purposes. To what extent, then, should the inventor of this technology be liable for its uses? In attempting to find a way to relate to this question, the Svea Court of Appeal concluded:

> In essence, these theories mean that an offense which involves a conscious risk-taking for a detrimental effect under certain conditions may still be allowed depending on the circumstances such as the nature of the danger, which

11. Day 8 in the first instance in the case against The Pirate Bay, Case nr B 13301-06. See Dagens Nyheter, 26 February 2009, http://www.dn.se/kultur-noje/rattegangen-mot-pirate-bay-dag-8/ [visited 23 August 2016].

> values the risks target, the social value of the offense, what precautions that have been possible and justified, etc. (Case B 4041-09, 26 November 2010, p. 24)[12]

The court then addressed this dilemma in terms of search engines, which is a function embedded in TPB, due to both YouTube and Google being mentioned by the defendants. A search engine that can assist in illegal distribution of content can still be legitimate, according to the court, but it has to be dominated by legitimate use to 'public benefit':

> If a search engine is characterised by primarily being a valuable tool in lawful activities and generally to public benefit, if this legitimate use dominates, but distribution or transmission of illegal material in spite of precautions cannot be ruled out, the operation of such a service, from an objective point of view, may be regarded as permitted in accordance with the aforementioned theories. (Case B 4041-09, 26 November 2010, p. 24)[13]

Should TPB be regarded as a search engine? And if this search engine is 'a valuable tool in lawful activities and to a general public benefit' then it would also be regarded as legitimate from a legal

12. Case B 4041-09, 26 November 2010, p. 24: 'I huvudsak innebär dessa teorier att en gärning som innefattar ett medvetet risktagande för en skadlig effekt under vissa förhållanden ändå kan vara tillåten med beaktande av sådana omständigheter som riskernas art, vilka värden riskerna riktar sig mot, själva gärningens sociala värde, vilka försiktighetsåtgärder som varit möjliga och befogade m.m'.

13. Case B 4041-09, 26 November 2010, p. 24: 'Om en söktjänst till sin karaktär är sådan att den i första hand är ett värdefullt verktyg i laglig verksamhet och allmänt samhällsnyttig, om denna legitima användning dominerar, men spridning eller överföring av olagligt material trots försiktighetsåtgärder inte kan uteslutas, kan driften av en sådan tjänst i objektivt hänseende komma att bedömas som tillåten med stöd av nyss nämnda teorier'.

point of view. References in the TPB Court of Appeal case were made to the Google search engine, on both sides. This judgment of intent was a central issue in the case. However, as Andreasson and Schollin (2011) point out, the requirements for establishing this intent to constitute contributory copyright infringement were significantly lowered in the Court of Appeal, while the Court addressed the debatable assessment of the 'social adequacy' of a service—that is, the Court lowered the standards for what is deemed as contributory infringement, which would then probably include any general search engine, but tried to balance up this vast criminalisation with an evaluation of whether the service is 'socially adequate', i.e. if it is of predominantly legal use as well as to the 'general public benefit'. This assessment is not easy to arrive at in a lot of cases and perhaps it is doubtful if it is even possible.

Peter Danowsky, legal counsel for the record industry association IFPI, described in his closing argument in the Court of Appeal Google as a 'pure search engine', implying that the TPB focus on torrent files makes it impure in terms of being a search engine (or service).[14] Note the metaphorical content of 'purity' here in connection with 'search engine' and the negative connotations that are embedded.

TPB as a bulletin board

Legal scholar Kristoffer Schollin was heard as an expert witness in the case, and stated, 'the best way to describe a tracker is that it is like a bulletin board'.[15] In the Court of Appeal, the prosecutor

14. In Court of Appeal, 12 October 2012, see Svenska Dagbladet, http://www.svd.se/naringsliv/digitalt/slutpladeringar-i-pirate-bay-rattegangen_5493311.svd [last visited 23 August 2016].

15. Day 8 in the first instance of the case against The Pirate Bay, Case nr B 13301-06. See Dagens Nyheter, 26 February 2009, http://www.dn.se/kultur-noje/

argued that TPB be seen as an 'electronic bulletin board'.[16] In this case, it would mean that the prosecuted would be seen as actively involved in copyright infringement if TPB were to be legally classified as an electronic bulletin board.[17] From a metaphor-theoretical approach, it forces us to apply what we know about the source domain (bulletin boards > electronic bulletin boards) to the target domain (TPB). The constant battle in court over TPB regarding the dichotomy of active/passive not only targets the characterisation of the actual functions of the website or the BitTorrent technology but also *the individuals* accused of being responsible for TPB. The prosecutor argued for legally supported metaphors that would frame the accused as playing an active role in supporting illegal file-sharing (contributory copyright infringement) and the defence arguments used legally supported metaphors that would frame the accused as passive or inactive.

TPB founders as entrepreneurs

Were the accused 'entrepreneurs'? How did the court regard this? Was it good or bad? Each metaphor is connected to specific values, of which some also are legally relevant. For example, if the individuals responsible are entrepreneurs, then a profit value is easily argued for. And the distinction of whether or not the activity is aimed at making a profit or not is very relevant to the calculation

rattegangen-mot-pirate-bay-dag-8/ [visited 23 August 2016]. In Swedish: 'Det bästa sättet att beskriva en tracker är som en anslagstavla'.

16. Day 7 in the Court of Appeal in the case against TPB, 12 October 2012, Case B 4041-09.

17. The so-called BBS-law (Act on Responsibility for Electronic Bulletin Boards, 1998:112) was adopted in 1998 and applies to services that electronically convey messages. The purpose of the Act is to establish responsibility for the provider to keep an eye on the bulletin board and to remove messages that clearly constitute incitement, hate speech, child pornography, unlawful depiction of violence, or messages where the posting user manifestly infringes on copyright.

of damages in the case. The District Court concludes that TPB has been run as a 'commercial project', which, as stated in preparatory legal works, indicates that a tougher sentencing is appropriate.[18] Palmås, a social scientist and entrepreneurship analyst, addresses the fact that the court in grave terminology described the cunning and ambition with which TPB operated as additionally incriminating, and concludes that these are the same type of innovative traits that we teach the students to take on the world with:

> The District Court's ruling suggests that it was not possible to demonstrate the financial benefit, but also listed other entrepreneurial traits of the accused as damaging factors. The TPB founders worked as a team, they investigated the appropriate organisational forms, and tried to think creatively about different sources of revenue—they had, in other words, practiced what we are today trying to teach to the thousands students of entrepreneurship at the country's universities and colleges. (Palmås, 2010, author's translation)

In short, the same characterisation that is especially attractive in innovation is here found to be especially incriminating. Another way to put it is that TPB made money from the online distribution of music and movies where the industry itself had failed to do so (cf. Palmås et al., 2014; Elert et al., 2016).

TPB as assemblage

Regarding the identity and existence of TPB, Fleischer has, for example, called it an assemblage (Fleischer, 2010). In his

18. Case nr B 13301-06, p. 79. Preparatory legal work prop. 1981/82:152, p. 18.

contribution to the Swedish anthology *Efter the Pirate Bay*, he explains that The Pirate Bay was 'a name of an assemblage of software, hardware, people, and symbols. No single component meant something by itself, but the explosive force arose from its reconciliation' (Fleischer, 2010, p. 264). The terminology has its theoretical roots in a French tradition that includes Gilles Deleuze and Felix Guattari (1987), and has been developed by Manuel DeLanda (2005). The point here is to present an alternative to seeing TPB as a seamless whole, to explicitly point out the multitude of functions sorted in the particular setting it consists of. The challenge that quite naturally follows from this is, what does it mean when these functions are rearranged at a later point? To what extent does a court case dealing with TPB in 2006 also regulate TPB in 2016?

Describing TPB in terms of an assemblage emphasises the notion of a construction put together consisting of different pieces, each with its own significance. And what has been assembled can also be disassembled, decentralised, and distributed. However, the legalisation of the activities TPB stands for demands a personalisation of the abstract or assembled entity that TPB is. The first grand problem of a networked activity is the one of picking the people most important to this networked activity. Even at this point, it is hard to draw the line. There are likely to be a large number of participants who have contributed, each in their own way. The roughly outlined categorisation of law does not easily fit into the great diversity of roles in a digital environment, where identity and types of contribution can differ considerably, and not even individuality needs to be an important distinction. In order to understand the organisational form of TPB, at least five main technological functions can be identified as necessary for a BitTorrent tracker site to work:

1) Domain name. The website needs to be found by its users, and have some sort of functioning interface.
2) Internet access. The website needs to be accessible through the Internet.
3) Search engine. Although many different categorisations of the torrents may be considered, a search engine is very helpful if the database is extensive. Which brings us to the . . .
4) . . . database of torrent files. Since no actual media content is found under the domain name and the actual webpage, the torrents that show exactly where this content can be found need to be gathered somewhere. It is these that are searchable. They, of course, need to be accessible.
5) Tracker. This feature communicates with the peers using the BitTorrent protocols, which can be termed the clients (this is however not used with *magnet links* [see subchapter 6.4]). Clients that have already begun downloading also communicate with the tracker periodically to negotiate with newer peers and provide statistics; however, after the initial reception of peer data, peer communication can continue without a tracker.

So, in short, there are three main functions behind the website—namely, the search engine, the torrent database, and the tracker that has to be connected to the Internet and tied to a domain name. It is these three main functions that have been associated with TPB in the court cases.

LIABILITY OF THE FUNCTIONS

In order to understand the future challenge for law here, it is important to understand that the 'assembled' functions can be disassembled and spread out over different legal jurisdictions and operated by a number of different subjects. This highlights the question of to what extent these parts of assemblages such as TPB are liable for what the whole assemblage leads to. In the aftermath of the first court verdict regarding TPB (1 April 2009) some of these other functions were also targeted by the same plaintiffs as in the TPB case. The responsibilities of the people providing the necessary building blocks are dependent on the legally accepted criteria for what the site is, or is not, doing. Number 2 concerns Internet access; on August 21, 2009, the District Court of Stockholm issued an interim injunction against Black Internet AB, stating that the company had contributed to copyright infringement by providing Internet access to the Pirate Bay file-sharing service.[19] A group of major movie and music companies from both Sweden and North America filed this suit against the broadband provider shortly after the court decision in the District Court Case against TPB. The step from passive contributor goes via the contributor becoming conscious of the nature of its contribution. Legally, this is how the passive contributor can become liable, because they have neglected the fact that their contributions are transformed into illegal actions. The result of the suit was that Black Internet stopped delivering Internet access for TPB, which for about a day went offline before it had rerouted its access. Black Internet appealed the District Court decision (3 November 2009) without success.[20]

19. T 7540-09, T 11712-09, 21 August 2009.
20. Case Ö 7131-09, Svea Court of Appeal, Black Internet, 21 May 2010.

Number 5, above, concerns the BitTorrent tracker. While filing the suit against Black Internet, the group of major American movie companies simultaneously filed a suit against ISP Portlane for offering services to BitTorrent tracker OpenBittorrent as well as against two of the men convicted in the TPB case, Fredrik Neij and Gottfrid Svartholm Warg, in order to prohibit them from participating in the operation of the file-sharing service (including the tracker). The law firm, MAQS, demanded that Portlane deny Internet access to the tracker function. The District Court did not, however, rule in favour of the American companies.[21] The District Court stated that according to the legislative history of the Swedish Copyright Act,[22] it is clear that more of an intermediary role is required than provision of Internet access for liability for contributory copyright violations. The Court's evaluation was that the application of copyright provisions in the manner the plaintiffs argued would mean too far-reaching responsibility for an ISP. They would be forced to exert control over what a tracker holder does and the extent to which the tracker was involved in copyright infringement or not. A significant distinction from the characterisation of TPB in this case, in relation to the characterisation of TPB in the original case, is that here the case regards TPB in its organisational form as of 2009— when the tracker function, at least, was no longer directly connected to TPB— while in 2005–2006 more functions were assembled under the same TPB flag. The court however ruled in favour of the movie companies regarding the two men convicted in the TPB case. The decision was appealed and the Court of Appeal upheld the decision against Neij and Svartholm Warg. It also overruled the District Court decision regarding Portlane and banned Portlane

21. Case T 17127-09, 1 December 2009.
22. Prop. 2004/05: 110.

from providing Internet access to a specific tracker (tracker. openbittorrent.com) involved in TPB.[23]

Fredrik Neij and Gottfrid Svartholm Warg, having just been convicted in District Court in the TPB case were, through the decision in District Court and later the Court of Appeal, prohibited from participating in the operation of the file-sharing service and risked a fine of EUR 58,000 if they were found doing so. However, later the same year (2009), TPB announced that it had closed its tracker down since the service now relied on so-called 'magnet links' that allowed peer-to-peer sharing without a central tracker to guide the swarm. Consequently (to sum up) instead of shutting down the activity that was assisted by the BitTorrent tracker, the parallel torrent technology developed into becoming independent enough to no longer rely on a tracker. This describes what Jonathan Zittrain (2006, 2008) has expressed as 'generativity'—that is, the possibility of restructuring and rebuilding that in many instances has been a characteristic of the digital coded medium. This is a main reason why this medium is so hard to control from a central point of view of, for example, law.

OUTLOOK: GENERATIVITY IN DECENTRALISATION

In order to concentrate on the development of p2p sharing, I return to the possible disassemblage. As can be seen from the differentiation into several necessary functions, none of these need to be located at the same place, under the same pirate flag, so to speak. They may be extremely decentralised in terms of

23. Cases Ö 8773-09 and Ö 10146-09, Svea Court of Appeal, Black Internet, 21 May 2010.

both geographical location and collective metaphors. This further complicates the characterisation of what a site such as TPB is, and it further asks rather intriguing questions in relation to an incredibly geographically and concept-dependent regulation of copyright. If a BitTorrent tracker site of the early TPB character were to be sued for complicity in copyright infringement, issues to be covered would include:

- In which country, and hence, under which legal jurisdiction?
- Which of the functions is illegal? Or is it the combination of
    - Search engine and interface?
    - Torrent database? (and)
    - BitTorrent tracker?

And then, what if the search engine is run by one party, the torrent database by another in another country, and the tracker function by a third party in yet another country? And what if the search engine not only finds torrents, but anything at all, along with the fact that the tracker aids in the sharing of a great deal of non-copyright-infringing material, and in fact the torrent database contains all kinds of both copyrighted and non-copyrighted material? The opportunities for decentralisation are all here. It could be said that BitTorrent sites are becoming meta-p2p—that is, decentralised catalysts for decentralised file-sharing. And, of course—with Zittrain's *generativity* in mind—there are other protocols that take over, or versions of the same protocol, that offer even more decentralisation, more of the nodes in communication with each other as well as being strongly encrypted, such as the 'magnet link'-distributed Hash Table (DHT) that first began to emerge in 2009. DHT may be described as a large ad hoc network of peers, passing on information requests about torrents without a central server, meaning no control or single point of failure. This means that no tracker function is necessary. By moving to a more decentralised system of handling tracking (DHT+PEX) and distribution of torrent files (magnet links), BitTorrent becomes even less centrally definable, meaning no central tracker that can be downloaded and no reliance on a single server that stores and distributes torrent files, and hence, no centrally located individuals to be held responsible.

NORMATIVE IMPLICATIONS OF SKEUOMORPHS

Which generalisable facts can we learn from the TPB case in terms of reusing conventional concepts in new circumstances, and what implications follow from judging the new with concepts from the past? For one, skeuomorphs come with normativity, and it is the past being normative in relation to the future. Already existing concepts and artefacts are constantly measuring the future and new phenomena. In this sense, we can never be free in our judgment of the new and abstract. We need concepts and we need to conceptualise the new. But what I argue in this chapter is that we need to be aware of this 'conceptual path dependence', particularly with regard to the powerful instruments of the law. We need to be aware of what Katherine N. Hayles (1999, p. 17) means when she refers to a skeuomorph as something that 'simultaneously focuses on the past and future, while reinforcing and undermining both'. It is this conceptually bound normativity that controls our actions and thinking when we immediately know how to use a digital camera, but at the same time are constrained in our actions by our understanding of it being based on and mimicking an analogue phenomenon from the past.[24]

The 'undermining' aspect of the skeuomorph lies in the fact that the digital phenomenon used by any of the legal skeuomorphs describes an analogue presence that exists within constraints that are significantly different from those of the digital phenomena. A parallel can be drawn to the use of the concept of 'copies' in copyright legislation, as outlined in the previous chapter; it means taking an earlier concept and reusing it in a new context under different constraints and conditions. The benefits are clear. We understand the new quickly, and the transition is in many aspects seamless and functional, e.g. we understand digital entities in terms of singular copies in a similar way to one physical copy being different from another physical copy. The downside, perhaps not as clearly expressed, relates to the fact that the conditions under which the skeuomorph exists are fundamentally different.

CONCLUSIONS

The TPB case is of interest from many angles—not only as a portrait of the obvious challenge to a legal construction, but also because it reveals some of the underlying dilemmas in the construction of this particular regulation in the digital context. Furthermore, the case clearly illuminates the dilemmas of applying a legal construction formulated to control making copies and their distribution to an environment that has so fundamentally

24. How norms function, especially social norms in relation to legal norms, has been studied and conceptualised in the discipline of sociology of law (cf. Svensson, 2013; and Hydén and Svensson, 2008).

revolutionised both the reproduction of copies as well as distribution. In fact, this case may be used to illustrate that parts of copyright have become metaphorical in the sense that the reality it attempts to regulate has changed essentially. The concepts have received new meaning, or perhaps more properly put, the concepts have colonised new practices and therefore *mean* more. In general, the key propositions of this chapter are:

1. Metaphors are fundamental to language and mind—abstract thinking is largely metaphorical;
2. Digital phenomena are largely understood through metaphors and skeuomorphs;
3. Law, in a digital society, is inevitably affected by this—concepts become skeuomorphs (their meaning changes over time, although retaining some connection to a [normative] past).

Skeuomorphs, thus, come with a price. Although they offer a quick understanding of new phenomena, they also to some extent hide the novel aspects of these new phenomena. In the TPB court case, the prosecutor argued for legally supported metaphors that would frame the accused as playing an active role in supporting illegal file-sharing (contributory copyright infringement), and the defence arguments used legally supported metaphors that would frame the accused as passive or inactive. By categorising the activities as either active or passive in relation to the ongoing, undisputed file-sharing somehow linked to TPB, the founders of TPB would be held liable or not liable for aiding any copyright infringement this file-sharing leads to.

Legally, the outcome of this battle of metaphors regarding the TPB makes all the difference. Speaking of skeuomorphs and the metaphorical practice of moving a concept from a source domain to a target domain, notice the extreme need for metaphors when naming and describing the coded entities and the digital organisation in this case: the 'torrents', the 'tracker', the search 'engine', peers in a 'network', 'links', 'magnet links', etc. And in the legal search for a metaphor that appropriately describes TPB, the unavoidable tools are inescapably more metaphors, such as TPB as a 'platform', 'bulletin board', search 'engine', etc. This means that metaphor is not incompatible with legal reasoning. Rather it is an inevitable part thereof and, one would suggest, a part that is of extra importance in terms of conceptual development in times of more rapid transition—for example, when a new technology is introduced. Law professor Bjerre (2005), for instance, concludes that 'the lesson for legal analysis is not to shun metaphor, or to seek liberation from it, but rather to realise that this aspect of thought is part of how the law functions, and that we can use it as an opening for reform though we must also live within its constraints'.

When it comes to TPB, given the possible rearrangements of the functions included in a site like TPB, it must not be forgotten that the original TPB case regarded TPB *as it was organised* in 2005–2006. The extent to which the case will function as a precedent for other cases and the extent to which the 2005–2006 version of TPB is in any way able to symbolise p2p file-sharing via the BitTorrent protocol when the very infrastructure has changed is also unclear. The form of organisation, the existence, the *is*, has changed, probably toward using even more of the inherent strengths of the Internet: heading for maximum decentralisation not only in its users, but also in its functionality of connection of users. The benefits of the verdict are therefore hard to find. If we view the question from the angle of intellectual property regulation, a number of studies have concluded that the weak support for upholding copyright law online is both well spread and intact after 2009. This means that four individuals have been punished for aiding a behaviour and normative precondition supported by a majority of the younger generation. The claimants, who won the court case, spent large sums on lawyers' fees for winning a case that only renders a right to damages that they most likely never will be able to retrieve and a hard to estimate bad-will amongst the young generation of file-sharers all around the globe. This should especially be seen in the light that the site itself has been operative all along, albeit with occasional downtime.

[7]

BETWEEN FORM AND FUNCTION IN (INTELLECTUAL) PROPERTY

The notion of property is at the core of our culture, often succinctly regulated in our constitutions. As the world and its social patterns and technological features have changed, the bond between the form and the function of property has been under constant negotiation. This is of particular relevance for the 'intellectual' and nonmaterial property in a digitised age. This chapter[1] deals with copyright regulation meeting the quite rapid societal changes associated with digitalisation, and it does so by reinterpreting Karl Renner's classical texts in the light of contemporary cognitive theory of conceptual metaphors and embodiment. From a cognitive theory perspective, I focus on the notion that the legal norms only *appear* to be unchanged—the Renner distinction between form and function. This includes social norms, technological development, and changes in social structures in general, which create a social and cognitive reinterpretation of law. This chapter, therefore, analyses the contemporary push for copyright *as property*, which I relate to historical claims for copyright as property as well as de facto legal revisions in intellectual property (IP) faced with the challenges of digitalisation. Of particular relevance here is what Renner described (in terms of

1. This chapter is a revised version of the article 'Karl Renner and (Intellectual) Property—How Cognitive Theory Can Enrich a Sociolegal Analysis of Contemporary Copyright', *Law & Society Review* 48(1) (2014c), 3–33.

Conceptions in the Code. Stefan Larsson.

property) as an 'institution of domination and control', and thus the increased measures for control that are added to a digital context in the name of copyright.

BETWEEN FORM AND FUNCTION

Karl Renner, the Austro-Marxist and early contributor to sociology of law who also became the first prime minister of the young Austrian nation, focused on the role of property and contract in changing Western European societies. In the early 20th century, he used the rather fierce metaphor of *Chronos* to describe the progression of property by stating that '[t]he evolution of property does not rest, it is like a Chronos who devours other people's children' (Renner, 2010 (1949), p. 110).[2] In this sense, I will also focus on the notion of property and its shifting claims over time. Influenced by the Marxist theories of his time, Karl Renner's theoretical examination includes a conception of society as a dialectics between a legal superstructure and an economic base (Grace and Wilkinson, 1978, p. 94). This means that Renner relies on a Marxist perspective to construct a sociological theory of law (Treviño, 2008, pp. 119–27). Of key interest in Renner's work is the analytical separation between legal institutions from their *social functions*—that is, in the words of Kahn-Freund (1949, p. 3), 'the factual results of their application'. Renner stresses that it is not law that drives the changes in its *substratum* when he argues, for example, that such 'economic change does not immediately and automatically bring about changes in the law' (1949, p. 252).

2. It is unclear if he thereby repeats a rather common confusion between the Titan Cronus/Cronos/Kronos—that devoured children—and Chronos, the personification of time in pre-Socratic society.

Since Renner so clearly addresses change, it is of particular interest how he conceptualises what it is that changes in relation to what he conceptualises in law and legal norms that do not change. Renner offers a view of how the form can remain constant even in transitional times, but at the same time, its content or function can be considerably modified. This chapter offers a contemporary and complementary perspective from the cognitive sciences on how the meaning of legal concepts can change with the context in which we shape their meaning. Renner's empirical study on the changing functions of property norms does, according to a commentator, demonstrate 'the significance of non-legislative developments in the law' and additionally identifies the potential for negotiating 'legislative involvement both before and after it does or does not take place' (McManus, 1978, p. 186).

RENNER AND (INTELLECTUAL) PROPERTY

Specifically, this chapter deals with copyright regulation meeting the rapid societal changes associated with digitalisation as a case of an unresting (intellectual) property making colonising claims over broadened fields of digital content. As mentioned, I do so by reinterpreting Karl Renner's classical texts in the light of contemporary cognitive theory of conceptual metaphors and embodiment. *Firstly,* I ask to what extent Karl Renner's theories on property can be used for an analysis of contemporary intellectual property; *secondly,* I ask in what sense IP is conceptually expanding as a concept in a digital context, displaying a need for conceptual metaphor theory for its analysis, and in what way such an analysis may be combined with and aided by Renner's theories; *thirdly*, I ask in what way the digital content is conceptualised as material and tangible objects and what this means; and *fourthly*, I ask, following Renner's argument on the institutions of control that follow property, to what extent this aspect of control is relevant also in the case of intellectual property in a digital society. The fourth point may very well be regarded as the most important here, in terms of how copyright and IP in many ways have been accentuated in the transition from regulating physical phenomena to also regulating digital equivalents. As I will explain further, it is crucial for the argument of this chapter to see how the digital environment in combination with the aspects of control that IP supports contribute, in fact, to collapsing traditionally supported distinctions between gaining access to a work, using it, and reproducing or copying it. Whether it relates to measurement of reading, shrewdly formulated as 'Your E-book is Reading You' in a *Wall Street Journal* article by Alexandra Alter (July 2012); or conceptualised in terms of Digital Rights Management and 'copy-locks'; or seen in the architectural settings of streaming services like Spotify, the development remains clear: copyright's claim is expanding in the digital domain. Although there are exciting methodological possibilities to be derived from, for example, the possible measurement of people's media use—an established tool in the 'data-driven' decision-making processes of Internet-based ventures and controlling the uses of digital artefacts—the focus here is on the aspect of control that this entails, which follows from a property-like copyright regulation operating in a relatively new dematerialised environment.

Although there has not been a continuously strong use or reinterpretation of Karl Renner's texts over the more than a hundred years since his first works were published, there have been revivals every now and then, particularly when *Die Rechtsinstitute des Privatrechts und ihre soziale Funktion* was translated into English in 1949 with an introduction by Kahn-Freund (Kann, 1950; Laski, 1950) and in the 1970s when a new edition was published (Auerbach, 1980; Bottomore and Goode, 1978; Kinsey, 1983; MacDonald, 1977; Robson, 1977; Shannon, 1977; cf. Treviño, 2010). Renner's ideas have also been cited in relation to participation and property rights (Leader, 1999) and, perhaps, particularly in relation to sociolegal research on land law (Whitehouse, 2010). In 1977, Peter Robson wrote a well-informed article on Renner stating that 'the ideas of Renner are still apt today in examining property and society. What has occurred has been changes in the appearance of property but consistency in its function as an institution of domination and control' (Robson, 1977, p. 221). With regards to property, contracts, and issues of control connected to them as a central function in society, it is rather uncomplicated to find a contemporary application for at least parts of Renner's work. In line with this, the debate on contemporary IP, particularly from an American perspective, seems to be increasingly focused on claims from the copyright industry that IP should become more property-like. This also follows on a broad trend related to increasingly consolidate copyright in a digital society to be an 'institution of domination and control', which I will return to below.

When Karl Renner provided us with the classical description of how property had been transformed over a period of time in its social function but not in its form, the role of property in a digitised world was, of course, not even imaginable. However, Renner's description is relevant for such an analysis, too. There are relevant parallels in contemporary society, I argue, to how '[c]hanges in society had successfully altered the form as well as the social function of property', as Robson (1977, p. 221) puts it. In addition, and as a driver for this chapter, I see a need for chiselling out more detail with the tools Renner provided us, and I therefore propose a complementary use of cognitive theory on conceptual metaphors in order to enable a more detailed study of the legal conceptual change at hand; when the letter of the law does not change, but its meaning does, this may be assessed by its dependence on context and societal relevance for its interpretation.

From a sociolegal perspective, as stated by many before me, it is of key importance to study and theorise changes in the meaning of legal language over time. This chapter demonstrates this dependency from the perspectives of language, cognitive science and conceptual change. The American lawyer David Mellinkoff writes that '[t]he law is a profession of words', emphasising the absolutely central role of language in law (1963:vi). This highlights the importance of understanding how language, meaning, and mind are constructed and also linked to the broader study of law, legislative change, and legal argument in relation to a social or societal context (cf. Amsterdam and Bruner, 2000). There are a number of studies that touch on cognitive theory in order to understand and analyse the legal fields, to explore, for example, how courts employ selective literalism (Tiersma and Solan, 2004), or how even blind people, in a conceptual sense, 'see race' because the understanding of race stems from interpersonal and institutional socialisations, and not a visual essence (Obasogie, 2010).

The development of law, as stated by several legal scholars, is generally conservative and therefore often retrospective. Embedded values are long lasting and consequent upon the main principle of predictability (Aubert, 1989, p. 62; Luhmann, 1972, p. 31ff; Peczenik, 1995, pp. 89–90). This has been described and analysed in terms of the 'path dependence' of law, mentioned in earlier chapters (cf. Larsson, 2011a; 2011c). Legal reasoning has its method in which categorisation is a key and inertia a virtue. As Steven Winter concludes:

> The structure of legal reasoning is essentially the same: it strives to reduce a complex problem to a policy, principle, propositional rule, or some other set of necessary and sufficient criteria. In theory, these definitional criteria will allow professionals to delineate legal categories with greater precision, draw appropriate distinctions, and then make correct decisions. (Winter, 2007, p. 870)

However, as both Winter (2001, 2007) and Mark Johnson (2007) have pointed out, this approach makes it hard to explain how law changes and adapts to new social circumstances. These issues have been addressed continually for many decades. Hohfeld, for example, complained that: 'Much of the difficulty, as regards legal terminology, arises from the fact that many of our words were originally applicable only to physical things; so that their use in connection with legal relations is, strictly speaking, figurative or fictional' (Hohfeld, 1913, p. 24). This addresses the difficulties of shaping and creating the language-based legal 'tool' that should on the one hand be predictable and reliable, and on the other hand is constantly reinterpreted in a changing societal context.

Karl Renner analysed the relationship between legal concepts of property and contract and patterns of social change in the development of capitalism in Western societies. His most important contribution was the aforementioned *The Institutions of Private Law and Their Social Functions* (2010/1949), first released in German in 1904 and translated into English by Agnes Schwarzschild, with an introduction and extensive comments by Otto Kahn-Freund, in 1949. Here Renner argues that law can adapt to changed social circumstances without necessarily changing its form or structure. In the words of Cotterrell, Renner argues that '[l]egal concepts can remain in the same form while fundamentally changing their social functions' (1992, p. 49), which has been labelled by Kahn-Freund as the 'functional transformation of the untransformed norm' (Renner, 2010/1949, p. 6). Renner concludes that the legal 'substratum' of property—'the social substructure'—had been completely revolutionised during the 19th century in Western Europe. It is against this fact that the legal setting must be displayed:

> Let us begin with this cardinal fact: the law of property has not changed. The Code Civil, the Prussian Land Law, the Austrian Civil Code and so forth—all these codifications which record the victory of the property norm, contain norms which are still valid today. The property norms of the new German code are even somewhat more strict than those of the earlier codifications. There has been no change of norms. (Renner, 2010/1949, p. 87)

The startling fact is that this occurs during a time of enormous economic and societal change, not the least due to industrialisation. How is it that the legal form can remain constant and yet regulate a society that is fundamentally different? Renner's analysis is relevant not only for his particular case but for legal conceptual development at large—perhaps, especially, in relation to when society undergoes rapid transformation. A year after the English translation of Renner's book was published Harold J. Laski reviewed it in a law journal (1950). In addition to celebrating O. Kahn-Freund's commentary, Laski addresses the change in legal norms in terms of their 'functional content' and 'inner essence':

> Since society is always dynamic, behind norms which often seem timeless there is infiltrated into the formal appearance a functional content which alters their inner essence at every turn of the road. (Laski, 1950, p. 390)

Without more thoroughly focusing on the ontological issues that arise regarding the inner essence of legal norms (which is done by Svensson, 2013, and Svensson and Larsson, 2012), our focus here lies in what we may see from a cognitive perspective, how changes in the language-based 'substratum' of law can be addressed. Laski explains, based on Renner's work, that 'legal norms are only apparently unchanged; at some time, they must either be interpreted in terms of purposes their makers never foresaw, still less desired, or they are overturned by those who can no longer accept what the original purposes do to the citizens of a society' (Laski, 1950, p. 390; see also Aubert, 1972, pp. 87f; MacDonald, 1977; and Shannon, 1977).

Due to the fact that 'law can adapt to change in ways that may not be readily apparent on the face of legal doctrine' (Cotterrell, 1992, p. 49), the proposed approach allows for a detailed study of the legal surface structures in explicit linguistic forms of expression that have the potential to reveal the underlying thought structures that govern a particular legal construction. It may thus show change in meaning where the specific concepts remain the same (Larsson, 2011b, pp. 131–32, 2013b). This, in my view, opens up paths toward a complementary, theoretical contribution drawn from findings in cognitive theory and, in particular, conceptual metaphor theory. The main point of connecting metaphor theory to legal analysis is to understand how the linguistic expressions and metaphors are linked with underlying conceptions and how our thinking thereby is framed and controlled by the metaphors that have become prevalent and which constrain or steer mental processes (Johnson, 2007; Winter, 2001). Here I thus argue that cognitive linguistics is significant to studies of the law. The important findings not only take into account the fact that metaphors play a much more fundamental role in mind and language than is traditionally acknowledged in theories of law (Johnson, 2007), but also concerns the framing aspects of conceptions and metaphors (cf. Lakoff, 2005). This is further emphasised by the process of *embodiment* of metaphors, and, hence, law—further outlined in chapter 2 and 3 above. In simple terms, law is in need of reification in order to be talked and thought about. This process, therefore, is of great interest to anyone concerned with understanding law's place in society as a cognitive, lingual, and cultural artefact.

THE P IN IP

From a cognitive theory perspective, I focus on the notion that legal norms only *appear* to be unchanged. In fact, their meaning may be in constant flux, but I contend that these changes need not be of a conscious character. Rather, we tend to live with the context that is changing the meaning. This context includes social norms, technological development, and changes in social structures in general that create a social and cognitive reinterpretation of law (Larsson, 2013c). This chapter, therefore, specifically deals with copyright regulation meeting the rapid societal changes associated with digitalisation. I will first address the contemporary trend of treating or arguing for IP as property and then continue to contrast this with a historical perspective.

In most jurisdictions, copyright owners have the exclusive right to exercise control over copying and other exploitation of their works. International treaties and directives focus on the control over reproduction of the protected creation. As outlined in chapter 5 above, on 'copies' the Berne Convention (Article 9) states that authors of literary and artistic works shall have 'the exclusive right of authorising the reproduction'; the Infosoc Directive (article 2) regulates 'the exclusive right to authorise or prohibit direct or indirect, temporary or permanent reproduction'; TRIPS (Art 14, section 2) states 'the right to authorize or prohibit the direct or indirect reproduction'.

In 1982, the famous pro-copyright lobbyist and Motion Picture Association of America (MPAA) president Jack Valenti argued in the U.S. Congress for equating 'creative property owners' with other property owners (Lessig, 2004, pp. 116f.). Already, the use of the term 'creative property' is a first step toward framing intellectual rights into a tangible goods property right. This is a mere example of how the pro-copyright industry makes property-based claims regarding the intangible goods that are being copied in the digital sphere because they are beneficial to this industry. To argue for equating copying with theft is yet another albeit simple way to colonise the digital phenomena with conceptions based in a physical environment. Herman (2008) shows that the notion of (tangible) property dominates the general mental image of copyright, and therefore much of the debate, resulting in a sort of pedagogical and rhetorical advantage (Yar, 2008) for those who propagate the conceptual links to the ownership of *physical* things. It then becomes an educational task of 'teaching' intellectual property when copyright, in fact, to some extent seems dysfunctional in a digital environment (Larsson, 2011b;

Larsson and Svensson, 2010; Svensson and Larsson, 2012). This is neither a completely external nor internal question for the law. For example, the statutory definition of copyright in the UK Copyright, Patent and Designs Act 1988, section 1, in fact, states that 'Copyright is a property right'.

Peter Jakobsson, a media and communication researcher, has analysed the contemporary shift of power from 'content providers' to what he calls the 'openness industry', including those players that benefit from lack of control on media distribution such as YouTube, Google, etc. Copyright regulation is central for his analysis:

> The for a long time dominant view that copyright is a limited monopoly—limited in time and in terms of the privileges that the copyright owner has in respect of the work—allegedly in recent decades has been replaced by a view that sees copyright and ownership as one and the same. (Jakobsson, 2012, p. 71)

This change of perspective is likely to have strengthened copyright holder positions (Lemley, 2005). According to Jakobsson, the increased use of the concept of intellectual property can be understood by the development of an increasingly neoliberal-influenced media politics (2012, p. 72), and Jakobsson argues that this is particularly true concerning the United States where the protection of private property has strong cultural roots. Even in Europe, it is probably easier to defend copyright by referring to an intuitive understanding of ownership than with abstract reasoning about time-limited monopoly (2012, pp. 71–72; see also Loughlan, 2007). The reasons, however, are not merely the outcome of conscious strategy or a rhetorical claim. For example, the study on the valuation of copyrighted content in the TPB case, presented in chapter 5, reveals a number of problematic assumptions that followed the click-by-click valuation used by the American complainants, a model that the Court approved. I show how a number of key assumptions are sprung from analogue conceptions of reality, and transferred into a digital context. This is a clear sign of a hard-to-detect, legal conceptual expansion of the meaning of 'copy' in copyright that does not 'fit with how the phenomenon is conceptualised by the younger generation of media consumers' (Larsson, 2013b, p. 615; see also Larsson and Hydén, 2010), which is of much greater general interest than one particular court case. The embodiment of abstract digital phenomena makes it deceptively easy to compare them to a notion of an already present phenomenon of physical copies, and uses a similar logic for how to deal with the digital equivalents. The problem is that this is a deception; they are not the same.

Copyright as property, historically

While there is indeed an increasing push to treat IP as property in contemporary digital society, this phenomenon is not entirely new. Strong property notions underpinned the early development of intellectual property; the push to use property notions to enhance the protection of rights holders seems in some sense to be cyclical and reactive. The reason can likely be found in the strengths that notions of property have—and have had throughout history. By framing copyright in terms of property, much is gained for those that hold the copyright.

The conception of copyright and the link to the notion of property has changed over the years both in terms of who ought to receive protection, for what reason, and for what type of creations. Historically, this becomes evident if we look at a time before the Romantic notion of the author grew strong, before the idea of the 'solitary genius' was established. For example, Schottenius Cullhed

(2012) has shown how the 4th-century poet Faltonia Betitia Proba has been differently regarded through the centuries. Proba wrote *Centos*, which refers to a method of composing, by which sentences and phrases are extracted from one or several texts and then put together in order to form a new text with a different meaning, resembling some kind of collage, remix, or assemblage today. Although positive responses to Proba's work can be found from the 8th century to the 17th, the perspective changes during the 19th and 20th centuries. Now this form of poetry is no longer considered 'real' literature, but is instead seen to be a disrespectful theft and misrepresentation of the originals. During the late-18th century, as Woodmansee (1984, 1994) has discussed, the modern idea of the author as a literary individual author emerges. The notion of an 'inspired genius' has played a part in strengthening copyright protection, which Hemmungs Wirtén (2004) shows through an analysis of the role Victor Hugo played in the establishment of the first international treaty on IP, the Berne convention.

However, historically, and one could argue today as well, there has been confusion about the role of authors and the industry benefits of strengthening authors' rights. Rose describes this as a 'contradiction between the romantic conception of authorship—the notion of the creative individual—that underlies copyright and the fact that most work in the entertainment industry is corporate rather than individual' (Rose, 1993, p. viii). Although Rose focuses on 18th-century Britain and 'literary property', his perspective is vital in exposing the origins of property. When the Statute of Anne was enacted in 1710, it was in part a legislative extension of a long-standing practice of the ancient London guild of printers and booksellers, the Stationers' Company (Rose, 1993, p. 4). An innovation in the Statute, however, was the limited term of protection, while the guild's protection was, or had been, perpetual. Authors of this time, in the early 1800s, were still very much dependent on patronage and writing was only to a very small extent an autonomous trade with its own economic strengths. The London booksellers sought to maintain their position by establishing that copyright was perpetual, despite the claims of the statute. While the booksellers' strength was still very much reliant on their intimate connection to, or appropriation of, the authors themselves, interestingly enough, their claims had a rights-based approach stemming from common-law rights of property transferred to them by the authors. These rights of property were dependent on the classical liberal discourse represented by John Locke's notion of the origins of property in acts of appropriation from the general state of nature (Rose, 1993, pp. 4f.). This meant an extension of the liberal theory of property, now targeting the work of the authors. This 'immaterial' property was here argued to be no less real and permanent than any other kind of estate. The confusion between the interests of the artists and those of the publisher, and the rhetorical use of this confusion, was early on exploited by publishers in the 18th-century version of copyright law in the United Kingdom. And when it came to controlling copyrighted works, the publishers gained strength from John Locke's theories on the rights following from property (cf. Volgsten, 2013, pp. 77f). The reason for this was that the alternative, a license solution, would be less beneficial to the publishers.

Even though property-based claims of copyright are not new, copyright itself has changed immensely over the years, particularly in its scope and reach. While it originally concerned authorship over books, it now also concerns music, architecture, software code, photography, etc. This displays the inevitable connection to the reproducing technology. As Eisenstein said before digitalisation entered the stage in *The Printing Press as an Agent of Change* (1980, p. 121): '[u]ntil it became possible to distinguish between composing a poem and reciting one, or writing a book and copying one; until books could be classified by something other than incipits; how could the modern game of books and authors be played?' Rose, too, notices the technological foundation of the regulation, and it is quite remarkable that even if Rose in the early 1990s had not yet seen the breadth of online creativity we witness today, he saw the construction of the institution of copyright as fundamentally challenged by new technology:

> Copyright developed as a consequence of printing technology's ability to produce large numbers of copies of a text quickly and cheaply. But present-day technology makes it virtually impossible to prevent people from making copies of almost any text—printed, musical, cinematic, computerized—rapidly and at a negligible cost. (Rose, 1993, p. 142)

Rose emphasises the role of technology in copyright's 'moral idea' in terms of its design having originated in 'printing technology, marketplace economics, and the classical liberal ownership individualism' (Rose, 1993, p. 142). This means that the benefits of the 'propertisation' of copyrighted goods have been around at least as long as the Gutenberg press. But how should the particular technological development from material to immaterial reproductions of copyrighted goods be regarded—as a change in degree or a change in kind? Renner may not be very helpful here, but in terms of what he calls the development of the social substratum, he claims that it 'knows evolution only, not revolution' (1949, p. 253)—that is, its development is a gradual process rather than the outcome of leaps. When it comes to the technological and ontological change that is relevant for IP in a digital context, I will return to this below in terms of 'control'.

Between legal and social norms

The gap between law and what can be termed social norms has, in the field of illegal file-sharing of computer programs, movies, and music via the Internet, been widely discussed (cf. Andersson Schwarz, 2013; Altschuller and Benbunan-Fich, 2009; Feldman and Nadler, 2006; Jensen, 2003; Larsson, 2011b, 2012; Larsson et al., 2012a, 2012b; Lessig, 1999, 2008; Moohr, 2003; Schultz, 2007; Strahilevitz, 2003a, 2003b; Svensson and Larsson, 2009, 2012; Tehranian, 2007; Wingrove, Korpas, and Weisz, 2010). Several studies have shown that a large segment of the global population sees illegal file-sharing via the Internet as a natural element of everyday life, irrespective of the intellectual property regulations of the state (Goodenough and Decker, 2008; Svensson and Larsson, 2012). In addition, or perhaps consequently, there is a counternarrative to the protectionist push. NGOs such as the *Electronic Frontier Foundation* (EFF) propagate digital rights and list what they consider patent abuses. The open source movement supports the use of open source licenses (such as Mozilla Firefox and Android), as opposed to traditional proprietary software (such as Microsoft Office). Furthermore, 'Creative Commons' is a well-known concept and movement in the copyright field and a good example of how writers, composers, photographers, and other creators can modify and oversee their copyright claims. On a more political level, there are a number of Pirate Parties in different countries; the Swedish Pirate Party, for example, received two European Parliament seats in the 2009 election, and the German Pirate Party received 9 percent of all votes in the 2011 regional Berlin Parliament election, and the Icelandic Pirate Party won 3 of the 63 seats in the national parliament in the 2013 election.

CONCEPTUAL LEGAL CHANGE

In addressing Renner's analytical approach to the difference between the legal form and its social function, intellectual property is of particular interest—especially, when trying to understand law in a digital society. The concept brings together all aspects contained in the argument put forward here regarding social change and its relation to law: it is connected to a particular language-based legal concept; it is central to most legal and economic systems; it is of substantial metaphoric content; and it is especially challenged by digitalisation (cf. Larsson, 2013c). For example, Mark Johnson states that 'we speak of intellectual property, such as ideas we have that can be copyrighted, patented, and excluded from use by others. Intellectual property is only metaphorically an entity, and it is only metaphorically transferrable to another for their use' (2007, p. 866). Since Renner does not clearly investigate what it is that, in fact, changes in the content of the legal norms that drive social change or change in law's substratum, there is a need here to express this type of change in terms of conceptual change, a change in meaning. One way to address this cognitive dimension of legal change is through conceptual metaphor theory.

As we have seen in previous chapters, conceptual metaphor theory has been used for analyses of legal processes or debates relevant to law in many studies. The concept of *metaphor cluster* is sometimes used to describe how concepts or lingual expressions can be bound together over a similar underlying conception, and thereby support the meaning of each other (Larsson, 2011b, pp. 60–1, 72–3, 2012b, 2013a), for example, to analyse metaphors in copyright and intellectual property (Loughlan, 2006).

If, for a moment, we view the concept of property from the perspective of categorisation, so central to human reasoning and language (Johnson, 2007; Lakoff, 1987; Lakoff and Johnson, 1999; Larsson, 2013a; Winter, 2001), then according to the classical view, which often is the prevalent one in law (Johnson, 2007), categories should be clearly defined, mutually exclusive, and collectively exhaustive (Lakoff and Johnson, 1999: 373–414). However, a cognitive approach following Lakoff and Johnson (1999) renounces the classic approach in that it accepts that natural categories are graded, tend to be indistinct at their boundaries, and inconsistent in the status of their constituent members. It is not that the classical view is entirely wrong, according to the cognitive approach; it is just that the categorisation based on shared properties only displays a small part of the story (Lakoff, 1987, p. 5). This means that even categories are to some extent culturally biased. Categorisation, for example, is expressed by Bjerre (1999) as a core activity even in law: 'Legal thought is, in essence, the process of categorisation . . . [c]ategorizing phenomena determines how they will be treated by the legal system'. Bjerre concludes that '[t]his basic truth is particularly important to the law because so much of it consists of arranging the world into language-based categories: "property", "contract", "good faith", "consent", "proximate cause," etc.' (Bjerre, 1999, p. 354). In terms of Renner, one could say that the categories of property and contract have remained, but their social functions have changed immensely (think of click-wrap agreements, for example). Bjerre's intention is to show that the concept of property, too, has a 'radial structure' (1999, p. 354), which Johnson describes in terms of its metaphorical content:

> The concept 'property' is not a classical category defined by a set of necessary and sufficient conditions. Instead, the concept is a vast, radially-structured category with a small

number of central members or prototypical cases surrounded at various distances by noncentral members, according to principles of extension such as conceptual metaphor and metonymy. (Johnson, 2007, p. 867)

Crucial to the conceptual change that claims IP as property lies the cognitive operation of *embodiment*. This borrowing of concepts from the physical and spatial, as well as the body, to make sense of abstract phenomena may very well play a role in the legal conceptual change of property into a digital domain. And, in terms of embodiment, there is constant borrowing in progress and interdependence on the surrounding context, the body as well as spatial relations, in order for language to become meaningful. In short, also these metaphors are based on our interaction with our physical and social environment. This is likely a process that makes it easier to speak and think of intellectual property in terms of physical things. It fits well with what Michael Reddy first identified as *the conduit metaphor* system (Reddy, 1979; see also Winter, 2001, pp. 52–6, 2007, p. 884), but it doesn't make it objectively true or disallow that other conceptualisations may be more fruitful in some cases.

CONCEPTUAL TRANSITION OF (INTELLECTUAL) PROPERTY

When Karl Renner analysed the changes in 'the substratum of law' in societal transition from a feudal to an industrialised society, he focused on the economic institutions in a Marxist fashion. Notwithstanding, he does not deny that the changes in social functions in the long run can lead to legal change (MacDonald, 1977, p. 8). However, when it comes to the legal substratum of property in a digital environment, this social substructure, I argue, could be analysed from the perspective of how concepts form and change their meaning from a cognitive point of view.

The combined metaphors of online piracy *as* theft in a copyright context very much relate to common ideas of (non-intellectual) property. As mentioned in chapter 5 above, Herman (2008) has analysed what he calls the metaphor of 'copyright is property' and hence the loan to the copyright debate of rights-based characteristics of the analogue, physical, and culturally well-founded ownership, especially in real estate (cf. Patry, 2009). The consequence of the rhetorical use of this metaphor is that it leads to a number of expressions that may seem meaningful within this particular property frame, such as 'trespassing', i.e. hacking technical barriers, and 'stealing' in the sense that they are copying or sharing computer files.

The property metaphor seems to dominate much of the legal debate and therefore also even frame the minds that seek to reconceptualise the problems that digital content offers. Metaphors highlight some aspects before others and therefor can frame complex and abstract issues or phenomena in a particular way, in this particular case resulting in a pedagogical and rhetorical advantage for those who propagate the conceptual links to the ownership of physical things. Which, in turn provides meaning, within this frame, to expressions such as 'theft', 'pirates', and 'trespassing'. Actions based on an analogue life of physical objects are metaphorically translated to speak for digital phenomena in a value-laden way. The metaphors, in this case, are therefore highly political and are conceptually controlling the legal argument.

When it comes to an analysis of the conceptual change that intellectual property and property has gone through, some extra attention needs to be focused on the origins of the concept of 'copyright'. Where English focuses on the 'reproduction' in *copyright*, Swedish and the Nordic countries stress the 'origin' in *Upphovsrätt*, and the French highlight the 'author' in *droit d'auteur.* When it comes to terminology concerning 'copyright', we also frequently encounter an anachronistic treatment. The concept, and the conceptualisation, has changed quite drastically over the centuries. For example, Hemmungs Wirtén studies the several centuries old historical origin of contemporary copyright and addresses the lingual differences in the times of the drafting of the 1886 Berne Convention for the Protection of Literary and Artistic Works, claiming that the Convention is a 'negotiation between legal systems, between copyright and *droit d'auteur,* between civil law and common-law traditions' (2011, p. 11, and as outlined in chapter 4). This means that there have been quite different conceptualisations and cultural origins and meanings that have had to be managed within similar legal concepts. Furthermore, what once concerned authorship and books has over the years been expanded to include composing and music, filmmaking and films, photography and photographs, and even computer programs.

When Rose (1993) and Woodmansee (1984, 1994) conducted their analyses over longer historical periods, they mainly relate to authorship, that is, in relation to the creation of texts and books. During the late 1700s, the rights came to include composers' printed music in English law (Fleischer, 2012, p. 122), which received no Swedish legal counterpart until nearly a century later (Volgsten, 2012). Another conceptual mash-up can be found in the legal difference between copyright protection per se and so-called neighbouring rights. As composers and writers receive protection as creators, so do musicians (performers) and record labels (producers of sound recordings) obtain rights relating to the specific recording. The origins of the rights to these different categories are different, but the categories are not always kept apart in public debate, which displays the blurry borders of the copyright concept.

Conceptual expansion of IP and the combo of Renner and cognitive theory

If, for a moment, we accept the notion 'copyright is property', or at least let the concept of property be blurred in terms of 'intellectual property', this conceptual expansion can be seen in the details of how the storage medium has changed from physical entities to digital. Whenever metaphors serve as conceptual bridges between one technology and another, it must be considered whether the norms that regulated the former phenomenon, which lends its name, can also affect or govern the new phenomenon (Larsson, 2013c). Cass and Lauer (2004) give the example of how the abstract and new digitised environment naturally requires concepts. Many of these are brought in from somewhat similar, but not identical, activities in the nondigital world. Cass and Lauer use the example of the transition between analogue photography and digital imagery. This metaphorical transition is likely often neglected in everyday life. Even if we were able to consciously detect the metaphors, the associations that are made instantly do not prioritise non-metaphors, as Glucksberg (2008) have shown. I have elsewhere (Larsson, 2011b) studied a number of concepts from the digital domain in order to shed some light on how copyrights concepts have expanded in a digital society (see Table 7.1).

The problem that emerges, then, is that whatever restraints and opportunities formed the characteristics of the source

Table 7.1 EXAMPLES OF CONCEPTUAL DEVELOPMENT FROM ANALOGUE TO DIGITAL. FROM LARSSON, 2011B

Source Domain		*Target Domain*
Analogue/Physical		Digital
mail	→	e-mail
trash can (garbage bin)	→	trash can (file deletion)
a copy (a record, a tape etc.)	→	a copy (.mp3, .jpg, .avi etc.)
theft (removing objects)	→	theft (copying digital files)
chat (casual conversation)	→	chat (digital instant messaging)

domain may not be the same for the target domain. In fact, the differences may be major. This is, as the third row of examples above might indicate, also applicable to copyright law and the objects it seeks to regulate. Before the days of digitised content, copyright law regulated reproduction and rights over the distribution of physical copies. That means that when a book was printed (in all the aspects of printing a book with covers, binding, ink, and glue) and distributed without the authorisation of the copyright holder, this action could be judged as a violation of the rights the rights-holder received from the law. The same applies if someone pressed vinyl records and distributed the music engraved into the plastic tracks. Today, the same regulations and the same legal concepts also regulate digitised content.

The technologically focused concept of 'path dependence', introduced in chapter 2 above, can be used to analyse copyright's development in a digital society in order to outline its lock-in effects (cf. Larsson, 2011a, 2011c). This would support the notion that 'conceptual path dependence' can explain how particular conceptions embedded in (copyright) law can become a conflicting social development in transitional times. Renner, too, observes the legally relevant conceptual path dependence in relation to societal change as a sort of mental inertia:

> The most surprising fact is the lack of social observation. Millions of people live among changing conditions, they daily feel their practical impacts, yet their theoretical implications do not become conscious to them. They think in concepts of a bygone generation. (1918, p. 51, translation in Aubert (1969))

This can be interpreted as an empirical approach to a legal assessment of how law is related to the conceptualisations and metaphors it relies upon. Even today, although we quickly adopt the new technologies of smart phones, emailing, and (for some of us) BitTorrent networks, it is hard to reconceptualise their meaning for law without a type of conceptual path dependency in the reuse of old concepts to understand the new (Larsson, 2013c, 2015). Arguably, therefore, even copyright in a digital society suffers from a 'lack of social observation'.

Copyrighted content as tangible and material objects

As law professor Winter explains, reification—the metaphorical making of abstracts into things—is a metaphorical process of great importance to law. For example, Winter claims that it is not possible to talk about law without the metaphor of 'object' (Winter 2001, p. 334). For a law to be broken, we must first conceptualise it as a thing that can be 'broken'. It must first be seen as an object that a criminal can 'take into his own hands'. In short, it is hard to imagine law without this reification (Winter 2001, p. 334), which, as mentioned, often is termed embodiment. This means that law in general seems to need metaphors that embody a physical, spatial or some sort of contextual source domain, as does copyright law in particular. In fact, much in the digital domain seems to need metaphors to be talked about or even thought of. With the material objectification of copyrighted content follows the meaningfulness of metaphors that are dependent on this conception. From this conception, I've argued (Larsson, 2012b), follows a pattern of lingual expressions of which the metaphor of copies is central. It asserts that the content can be replicated in exact identical packages, or copies, originating from an original source. These copies can then be owned, replicated, stolen, and pirated, which in other words means that they can be clustered according to a certain pattern that collectively describes the underlying conception. Loughlan speaks of 'metaphor clusters' in intellectual property debate and analyses several clusters she identified:

> The first metaphor cluster draws upon some highly negative images of lawlessness, and violent, predatory behaviour (pirates, predators), exercised against helpless victims; or of a creature eating away at and undermining the health and well-being of innocent victims (parasites); or a thief who by stealth removes what is not his or hers from an innocent owner (poachers); or a person riding for free while others must pay (free-riders). These metaphors occur both by themselves and, frequently, together, compounding the negative effect of each metaphor. (Loughlan, 2006, p. 217)

In relation to (nonintellectual) property and the norms surrounding it, theft is one of the clearest breaches. It can be described in the words of sociologist of law Vilhelm Aubert: '[a]mongst those legal rules that protect the position of the owner, the regulations of theft are one of the most simple, most stable and most known' (Aubert, 1972, p. 91, my translation). The example of 'stealing' in relation to 'sharing' in a digital context can be used to illustrate a type of 'battle of conceptions'. What from an analogue perspective is regarded as theft (an action with highly negative connotations) is from a digital perspective regarded as something else, with less or no negative connotations. Normatively, it could be said that these actions are not comparable. The legal concept of theft is closely related to the conception connected to 'copyright as property', and describes how the idea of property rights is formed in an analogue reality and transferred to a digital one, which is a process containing a number of great challenges (see Loughlan 2007 on 'theft' and intellectual property). One could also argue that not even the piracy motto of 'sharing is caring' is as (empirically) meaningful anymore. In a streaming-based access to culture, the copies themselves are not likely to be fully conceptualised, let alone the notion of owning or possessing copies, but these have more likely surrendered to the notion of having access to an archive and the content when wished for.

However, an example of the rhetoric surrounding theft, as well as a case in which Internet Service Providers (ISP) are seen as having a key role in copyright enforcement, can be found in the deal that was struck in July 2011 between a coalition of entertainment industry groups and several major US Internet providers to fight online infringement. The deal resulted in a 'Copyright Alert System' that was launched February 25, 2013.

The key idea is to notify and 'educate consumers about the importance of copyright'.[3] Cary Sherman of the Recording Industry Association of America (RIAA) commented on the deal by stating, '[t]his groundbreaking agreement ushers in a new day and a fresh approach to addressing the digital theft of copyrighted works' (*Wired*, July 7, 2011). The problem of arguing that file sharing is theft lies, of course, in the aspect of loss. There is no loss when something is copied, or the loss is radically different from losing, say for instance, your bike (cf. Yu, 2011). The loss lies in that you are likely to lose someone as a *potential* buyer of your product. The 'theft' argument is an example of how a conception tied to a traditional analogue context is transferred to a newer, digital context (Larsson, 2012b; Larsson and Hydén, 2010). It describes a change in the substratum of property, without changing the letter of the law, in the sense that '[l]egal change (no less than stability) is contingent on, and therefore constrained by, the social practices and forms of life that give law its shape and meaning' (Winter, 2007, p. 897).

KÖRPERLICH AND CONTROL

Although the explicit focus in this chapter regards the underlying understanding of what property is in terms of an expanding notion of intellectual property, this image has to be complemented with actual and explicit legal change as well. Even though I specifically address the 'contextual' shift that affects the interpretation of legal norms, it is a fact that copyright regulation—whether through treaty negotiations in the name of trade, through EU Directives, or through national lawmaking that is

3. www.copyrightinformation.org [last visited 24 August 2016].

more or less affected by the intervention of strong pro-copyright lobbyists—has undergone a quite dramatic change over the last thirty years or so. Therefore, in this section, I firstly address the conceptual and contextual aspects of the changes and secondly the explicit legal manoeuvring that has been conducted in this field. All under the notion of *control*, which I argue is of particular interest in a digital context.

Lakoff and Johnson (1999) claim that the mind is not only 'corporeal' but also passionate, desiring, and social. They emphasise the meaning of the body in the world, its connectedness to it, and describe it in terms of 'the properties of mind are not purely mental: They are shaped in crucial ways by the body and brain and how the body can function in everyday life' (1999, p. 565). They clearly state, in terms of embodiment, the conceptual dependence on 'what we walk on, sit on, touch, taste, smell, see, breathe, and move within. Our corporeality is part of the corporeality of the world' (1999, p. 565). It means that how we understand any abstract phenomena, including ownership over digital entities, is stained by models from a physical, spatial and 'corporeal' frame of understanding.

Renner uses the term 'corporeal' on a number of occasions to describe the transformation that property has gone through and that it now is 'no longer burdened with physical substance' (2010, p. 217). It was the translator Schwarzschild who chose the word 'corporeal', in translation from the German 'körperlich' (see pp. 81, 85, 89, 104, 107, 113, 217, 278, 293):

> We see that the right of ownership thus assumes a new social function. Without any change in the norm, below the threshold of collective consciousness, a de facto right is added to the personal absolute domination over a corporeal thing. This right is not based upon a special legal provision. It is

> the power of control, the power to issue commands and to enforce them. (Renner, 2010, p. 107; see Aubert, 1969, p. 34)

This is a process that can be seen also in the expanding claims of contemporary intellectual property that increasingly attempt to see 'copyright as property' in terms of a physical object. This reification could be described in Renner's words: 'Legal property is a corporeal object' (Renner, 2010, p. 278). An important bridge in the conceptual claims of 'copyright as property' lies in the aspects of control that have increasingly been added to the digital environment. Renner sees that the added aspect of control as matter-related only changes to 'control of human beings, of the wage labourers, as soon as property has developed into capital' (in Aubert, 1969, p. 34.). In a digital domain, control is essential, too, for intellectual property to be upheld. One of the main consequences of the strong path dependence of copyright is that legal enforcement is also experiencing important changes when it comes to the opportunities offered by tracking our digital traces, to *control* the flow of the Internet, and to track the identities that breach the laws of intellectual property. More individual traffic data are stored, the data are stored for a longer period of time, and accessibility to the data is made easier for not only policing entities but also rights holders in order to map and identify infringers of copyright. This means a potential for new ways of legal enforcement and mass surveillance over the multitude of habits and secrets of our everyday lives (cf. Larsson, 2011c, p. 30).

From a cognitive point of view, these surveillance-like consequences of an increased push for copyright enforcement in a digital context are of particular interest. To what extent are we, for example, adapting our behaviour in the awareness that we might be traceable in our online activities? What does it mean that the digital traces we leave behind in terms of traffic data, social media posts, and search engine patterns offer well-functioning profiling for those who can aggregate the data? (in terms of privacy, see Halbert and Larsson, 2015). Inspired by Foucault (1991), who was inspired by Bentham's prison design, some scholars discuss the notion of *panspectrism* in relation to the increasingly networked, logged, and digitalised way of life we lead and the contemporary possibilities for surveillance in terms of how the 'supervisor' can see more than is possible in a *panoptic* version of surveillance (Andersson Schwarz and Palmås, 2013; cf. Kullenberg, 2010; Palmås, 2009; Vaidhyanathan, 2015). To what extent, then, would it be meaningful to view copyright from Renner's perspective of control as a result of becoming (regarded as) property—and then capital?

Changes in legal norms in a digital era

As mentioned, there has been substantial legal manoeuvring in intellectual property regulation during the last few decades related to digital development; thus, I will mention a few of the most important examples below. The big shift on a global level for copyright regulation has concerned its metamorphosis into trade regulation. This goes hand in hand with a protectionist approach to both the scope of the regulation as well as the length of time that the protection is offered. The copyright historian Rose (1993) acknowledges as a key issue in his investigation of the notion of the author as something inseparable from the commodification of literature. This, I argue, plays a key role also in understanding the relatively contemporary development of copyright and intellectual property into a primarily trade-related legal construct in a global arena. Peter Drahos and John Braithwaite bear evident witness of this in *Information Feudalism* (2002). The development of control mechanisms, as well as the propertisation of copyright in a digital era, need to be understood in relation to a combination of the inherent ontological shifts that digitalisation offers, which writers like Lessig and Zittrain address, and the commodification of cultural expressions caused by the shift in copyright to become a global trade-issue, on which writers like Drahos and Braithwaite focus. While the latter may explain why some of the stalemate exists, e.g. its path dependence, the former displays the seriousness of the matter.

In *Information Feudalism* (2002), Drahos and Braithwaite show how global intellectual property is subordinated to trading interests, notably the TRIPS Agreement, which appears to be designed by a small group of US-affiliated industry stakeholders at the helm (2002, pp. 10f.). TRIPS is, of course, interesting from a copyright perspective, because it sets up copyright minimum standards for the majority of members of the World Trade Organization (WTO). This means that although the traditional arguments for a stronger copyright are there in the shape of protecting authors and artists, the content of the legislation is focused on industrial concerns:

> Beneath the dissembling rhetoric about the need to protect authors and provide incentives lay a harsh global economic reality of cartelized publishing industry, price fixing, and world-market agreements. (Drahos and Braithwaite, 2002, p. 76)

Drahos and Braithwaite show the face of the industry and lobbyism and the entanglement in law-making on a global level. This has also been evident in the secrecy surrounding the *Anti-Counterfeit Trade Agreement* (ACTA), which was negotiated between a dozen nations' representatives during 2010–2011 (Larsson, 2011c), receiving substantial critique (Geiger, 2012). The method of secrecy as a way to effectively change national Intellectual Property legislation by circumventing every democratic concern seems to have been reproduced in the Trans-Pacific Partnership (TPP)—finally signed on 4 February 2016, after seven years of negotiations—where the United States continuously acted as a strong pro-protectionist enforcer (cf. Patry, 2012).

At an EU level, the so-called *IPR Enforcement Directive* (2004/48/EC) was approved by the European Parliament on March 9, 2004. This Directive, which was implemented in national legislations, was aimed at making it easier for the copyright holders and their representatives to extract identification data from the Internet Service Providers when an IP-number was suspected to have been involved in infringement in any IP. A Swedish study concluded that the legal change did not make the younger generation feel that it was morally wrong to break copyright law in regards to file-sharing copyrighted content. The social norm was equally weak before and after the directive was implemented (Svensson and Larsson, 2012). When it comes to national legislations of particular interest, the British *Digital Economy Act* of 2010, the French *Haute Autorité Pour la Diffusion des œuvres et la Protection des Droits sur Internet* (HADOPI), and the U.S. *Digital Millennium Copyright Act* (DMCA) stand out. The first two relate very much to traceability in the digital context by pushing for obligations for ISPs regarding the traffic they mediate, as in the aforementioned deal between the US content industry and ISPs to create a 'Copyright Alert System'. The ISPs are expected to collaborate with copyright owners, to identify infringers, send notifications to alleged infringers, and keep copyright infringement lists (Mendis, 2013). The Digital Economy Act shares with the HADOPI the threat that disconnection from the Internet can follow after three warnings. This rather disproportionate punishment is part of the law despite the fact that the Digital Economy Act 'is unlikely to succeed in its central purpose to control unauthorised digital copying' (Mendis, 2013, p. 60). The U.S. Congress passed the DMCA in 1998 and it strengthened copyright in a number of ways, of which Lessig regards the 'anticircumvention' provision as 'particularly troubling' (Lessig, 2002, p. 188). The DMCA includes a regulation of devices that are designated to circumvent copyright protection; this regulation is also found in the *InfoSoc Directive*. The DMCA has been criticised for disrupting the balance between content owners and the public. According to Vaidhyanathan (2001), the DMCA results in that 'content providers can set the terms for access to and use of a work. There is no balance if the copyright owner has all the power' (2001, p. 159). The 'anticircumvention' is part of what sometimes is called *Digital Rights Management* (DRM). Even when DRM is, in practice, failing, it has been strongly promoted by a pro-copyright industrial discourse. This has been seen by critics as a way to authorise copyright to become even more powerful than ownership, for example, in terms of consumers buying music CDs yet still being restricted in what they are allowed to do with the purchased product (Gillespie, 2007).

Why IP is reaching further in the digital context

One of the aspects of a far-reaching copyright regime in a digital context is that it, in fact, reaches further than traditional property rights. The copyright holder can claim stronger control over the ways in which products should be used than the producers of, for example, a chair can. Vaidhyanathan (when he analyses DMCA) describes this in terms of an erosion of the first sale doctrine (in US law):

> When a work is sold, the copyright holder relinquishes 'exclusive' rights over it yet retains 'limited' rights, such as restricting copying or public performance. But under the first sale doctrine, the consumer can highlight a book, copy portions for private, noncommercial use, resell it to someone, lend it to someone, or tear it up, without asking permission from the copyright holder. Because the DMCA allows content providers to regulate access and use, they can set all the terms of use. (Vaidhyanathan, 2001, p. 175)

From a conceptual point of view, digitisation means that the addressed area of copyright law has increased, and it is an ontological change; the works do not exist in the same way that they used to exist. The storage devices used to be physical during the era that created what Lessig (2008) calls the 'Read Only culture', that is, where production was separated from those who consumed. In a digital context, this type of 'copy control' is harder to maintain, which has to a high extent boosted what Lessig consequently calls the 'Read/Write culture' and has made policy makers respond with expansion. This ontological shift of the protected works is of key relevance when analysing the shift that copyright has undergone in a digital society. Perhaps, Vaidhyanathan (2001, p. 152) says it the most lucidly:

> The digital moment has also collapsed the distinctions among three formerly distinct processes: gaining access to a work; using (we used to call it 'reading') a work; and copying

> a work. In the digital environment, one cannot gain access to a new story without making several copies of it.

This is also what Lessig has found to be an exaggerated focus on copies: '[f]or while it may be obvious that in the world before the Internet, copies were the obvious trigger for copyright law, upon reflection, it should be obvious that in the world with the Internet, copies should not be the trigger for copyright law' (Lessig, 2004, p. 140). Vaidhyanathan states that this focus has lead the policymakers to a troublesome choice:

> Now that the distinctions among accessing, using, and copying have collapsed, copyright policy makers have found themselves faced with what seems to be a difficult choice: either relinquish some control over copying or expand copyright to regulate access and use, despite the chilling effect this might have on creativity, community, and democracy. (Vaidhyanathan 2001, pp. 152–53)

This sheds some light on what it is that is conceptually new and particularly troublesome in the digital context when it is compared to the more than century-old legislative conception of copyright as linked to property.

The issue of control, therefore, seems to be the key battle in the copyright war. Renner comments on the aspects of control that follow with the institution of property:

> Supervision is delegated to special functionaries, and thus relations of superordination and subordination are made into an organic whole. Thus the institution of property leads

> automatically to an organization similar to the state. Power over matter begets personal power. (Renner 2010, p. 107)

With control follows the type of 'architectural' imperatives for behaviour that can be found in the particular setting in which any transaction takes. Here, one could speak of 'code as law' in the sense that Lessig claims, in which the digital domain, too, is subordinate to the exact conditions that the (programming) code allows (1999, 2006). Where Lessig's object of analysis is the digital code in relation to regulation, Renner's object of analysis is property, for example, related to production and factories, and regulation:

> Wage labour is a relation of autocracy with all the characteristics of despotism. The factory is an establishment with its own code with all the characteristics of a legal code. It contains norms of every description, not excluding criminal law, and it establishes special organs and jurisdiction. (Renner, 2010, p. 114)

This, of course, is one of the reasons Renner has been seen as an inspiration within the sociology of law field: this idea that not only state law is the important, normative imperative, but also that the factory is an establishment with its own code. This is also the reason Lessig's account dealing with programming code *as* law is valuable to those working on sociolegal topics, in relation to the Internet and the digital development.

I have elsewhere elaborated upon how the digital moment has changed our perception of what 'should' be protected and that the law tries to 'artificially' maintain constraints that were lost when moving from analogous storage devices to digital. This, in other words, is of key interest for identifying how people perceive the legitimacy or illegitimacy of the legal construction in light of its conceptual expansion (Larsson, 2013b, p. 631).

SUMMING UP: CONCEPTUAL LEGAL CHANGE

The key focus here has been to study conceptual legal change by adding a cognitive, theoretical perspective to address the same question that Karl Renner addressed: 'How is it that the law of property can remain unchanged and still function in a very much changed society?' Recognition of metaphorical thought, and the methods of conceptual analysis, demonstrates how legislative statutes express significant aspects of our social reality that cannot be devalued by reductive approaches to legal reasoning. Although the meaning is very much bound to specific patterns, these patterns can, and probably often do, differ from the 'objectively' defined patterns of meaning. In terms of property law and the way in which IP has developed in a digital society, it seems that core concepts to a high extent remain the same, in the same form, but have altered in their social function or substratum, to use Renner's terminology. This also goes for a number of underlying conceptions of control over copies, more protection, an increase of control functions—which, in parallel to Renner's analysis, seem to follow with the propertisation of copyright. Albeit the legal construction is in a sense similar from before, the regulatory claims have expanded—due to that a complete new set of (digital) actions and phenomena have emerged to be claimed—without a change in terminology. Of particular interest here, and perhaps as an indicator on propertisation and substratum-change, is what Renner described as an 'institution of dominion and control' tied to property. At the same time, we may remind ourselves that although there is a contemporary trend pushing for the notion of 'copyright as property', these arguments were not unheard of through the ages before digitalisation. The notions of property, which legitimises ownership and control, may here be seen as a conceptual tool for copyright holders to argumentatively strive for, in order to gain support for restraining cultural flows and public uses of texts, music, and movies—unless these flows are controlled by copyright-holding industries. One can here also notice that the trend of colonising more phenomena, becoming stronger and more protectionist in a digital society, is also affecting other legal fields (for example concerning privacy)—which is in line with Renner's intense metaphor of Chronos, where property—like Chronos—'devours other people's children'.

[8]

CONCLUSIONS

Conceptions in the Code

In this book, I have presented research showing how metaphorical thought is fundamental to our understanding, use, and construction of abstract phenomena—and thereby how dependent we are on metaphors in general. I have mainly related them to the contexts of law and digitalisation—two vastly important phenomena that depend on abstractions—with the intention of analysing some of the legal challenges linked to technologically related social change. There are *conceptions* in, and underlying, both law and digital architecture—that is, in *the code*. One important aspect of this is the role of courts, as Blavin and Cohen have put it:

> When courts encounter new technologies not yet anticipated by the law, their reliance on analogical reasoning plays a profoundly important role in the application of proper legal rules. Courts, however, have demonstrated a bad track record in adopting the appropriate analogies or metaphors for these new technologies. (Blavin and Cohen, 2002, p. 267)

The court's role is vital, as seen in chapters 5 and 6, respectively, on 'copies' and on the TPB court case, but the purpose of this book has been even broader, in the effort to better understand how law meets new technology and the new phenomena it creates. By pointing out how we understand abstract phenomena

Conceptions in the Code. Stefan Larsson.

metaphorically, I have attempted to unravel some of the legal challenges that we have to tangle with as a result of rapid technology-related change.

This concluding chapter is divided into six main sections in which I begin by briefly summarising the points made in the book and then, in *the second section*, focus the implications of using already established concepts to understand technology-related phenomena—for example, by using skeuomorphs and what I call conceptual path dependence. The *third section* focuses specifically on copyright in relation to the changing, and sometimes challenging, conditions that follow in the wake of digitalisation. The *fourth section* explores the regulating function that imagery has and links this to power—which means that certain actors are more inclined to control the prevailing imagery, or framings, than others. In *the fifth section* I focus on a more forward-looking perspective in addressing the metaphor of 'open' as a central metaphor for how new media, digital architecture, and regulatory models navigate in a contemporary discourse on digital innovation. This includes a speculative account on the role of 'liquidity' as a way to address conceptual turmoil in times of change. In the concluding *sixth section*, I aim for a synthesis with regard to science's objectives concerning the normative dimensions of technology-related social change.

METAPHORS, LAW, AND DIGITALITY

As mentioned previously, cognitive scientists have demonstrated that metaphors are necessary for how we think, and metaphorical thought provides us with imagery that allows us to speak about abstract phenomena. This is deeply rooted in our language and minds. Furthermore, *conceptual* metaphors are, simply put, the actual building blocks of our reference-bound thought. However, these building blocks are not empty of values. Underlying conceptualisations frame thought structures—or, differently put, structure our thoughts—and can either emphasise or downplay and conceal the functional aspects of a phenomenon. That is to say, they influence whether we understand creativity as singular genius or a social manifestation; music stored digitally as a demarcated packet or as a flow; relations as a journey; prostitution as either a commercial transaction or violent abuse; unlawful copying as theft or cultural achievement. Consequently, it can be problematic that we generally lack awareness of the power that metaphors and concepts have and that they tend to remain unobtrusive and are taken for granted in our language and how we understand the world, without being clearly expressed in our conscious daily lives.

In relation to law, I conclude that law, too, is highly dependent on concepts and metaphors to be communicated, understood, and applied. Law is essentially concept-dependent while simultaneously being conservative and slow to adapt. This combination leads to a sort of path dependence (Larsson, 2011c, 2015). However, there is nothing inherently problematic about using a historised way of approaching new technology. On the contrary, in many situations, one of law's strengths is that it is *not* a political weathervane, and can resist powerful lobby attempts to control it for their interests. The problems arise when law is conservative in the sense that it loses the ability to conform to new ways to organise society, new ways of concluding agreements, new forms of labour, new forms of property ownership, new ways of creating, and new ways of participating in these cultural expressions. It may then continuously protect the interest of the powerful that was inserted during a more formative period in history, under conditions that may not exist anymore. As society and social norms change, law risks becoming a political distributor of power rather than a legitimate representative of what is perceived as right and wrong, in holding on too strictly to past values.

I also develop the experiential dependence that cognitive scientists refer to as *embodiment*, partly in order to demonstrate the link to material conditions. How we understand things depends to a large degree on our experiences of space, of spatial relations, bodies, and surfaces. I argue that since our embodied experiences would also appear to control our understanding of digital processes—in the context of digitalisation and the legal challenges it poses—conceptual metaphors become highly interesting to study. These changes in our material conditions—how communication works, how our behaviour leaves masses of quantifiable digital traces in a manner not previously experienced—affect our understanding of our reality, while also providing science with new understandings of us.

When Lakoff and Johnson originally developed the metaphor theory I have, to a large degree, allowed to permeate this book's analyses, the first thought that struck them was likely not what this would entail for law and the legal system. And they almost certainly did not envision that a future digital network would demand a wave of new metaphors, or that this network would also challenge legal concepts. As mentioned in the introduction to this book, it is thereby possible to state that without concepts and imagery, we cannot speak about or understand the new, and without metaphors we have no concepts for new phenomena. And in relation to the highly digitised times we live in, consider the following metaphors: The cloud, networks, streaming, dataflow, desktop, interface. Or new phenomena that resemble old phenomena: Email, digital photos, digital copies, and eBooks. We conceive of them partly as something that is already established, and partly as something new. 'Understanding' should therefore not here be interpreted to mean that we understand things *as they are*, but rather as one of potentially several *ways of understanding*, as part in a cultural negotiation depending on both aspects of language and technological conditions, along with our spatial and bodily circumstances.

CONCEPTUAL PATH DEPENDENCE

Beginning more generally with the legal system and laws in times of rapid technological change, one could argue from a cognitive-theoretical perspective that legal norms only *appear* to remain constant. Having established that point, one could then argue that they are, in fact, in constant movement, although the changes may be hard to discern and quantify. Their interpretation is directly linked to the social and societal context they exist within, which is dependent on changes in social structures, cultural expressions, and technological possibilities. It is the contextual change that produces a reinterpretation of legal concepts that seem unchanged in their form, but which can have a very different effect or function, leading to what McManus (1978, p. 186) has called 'non-legislative developments in the law'. In the previous chapter, I develop precisely that point, particularly in relation to ownership and control. I use an example that has become something of a classic in sociology of law, namely Karl Renner's study on property and contract law during the industrialisation of Western Europe during the late 1800s (Renner, 1949). Renner's theories focus to a great extent on the difference between form and function, and concludes that the legal property form and how the concept is worded remains largely unchanged—despite society, and therefore the application of the law, having changed fundamentally as a consequence of industrialisation. How is law to function without large revisions and adaptations to new conditions? The chapter uses Renner's theory in combination with conceptual metaphor theory to further understand the changes and development of property rights in a digital and immaterial world.

Somewhere along the process, law can change almost unperceivably when the interpretation of its concepts come to include new phenomena, while its relation to the phenomena it was originally intended to regulate simultaneously may grow more strained and distant. Having concluded that legislation which emerged largely in a pre-digital era is encountering difficulties in regulating primarily digitally mediated phenomena satisfactorily, it would seem reasonable to seek the explanation from a chronological perspective—i.e. that the thought structures that underlie a specific law or paragraph are quite simply anachronistic in a changing world. These may have arisen as a result of inherent legal and conceptual lock-in effects that shape our previously mentioned path dependence. As explained above, there are different fields where this theoretical figure has had particular explanatory relevance, such as in spatial planning and in technological design, but my focus is primarily on what I call *conceptual path dependence*; i.e. when different *ways of understanding* certain phenomena become subject to 'lock-in effects'. On the one hand, this is unavoidable since a completely ahistorical interpretation of the new would be without reference, incommunicable, and without meaning. We indeed require established concepts to understand the new. To some degree, we are all products of our cultural backgrounds, biological and environmental prerequisites, our languages, and how we conventionally tend to interpret phenomena.

When computer scientists Kimberly Cass and Thomas Lauer studied email and digital photos, they did it with a focus on how these phenomena depend conceptually on somewhat different, yet similar, earlier phenomena. Digital photography and the digital camera are similar to analogue photography and the analogue camera in many ways. The advantage of building a conceptual bridge between them is that one can quickly form an understanding of the (digitalised) phenomena by basing it in the old (analogue). Thus the new becomes framed within an interpretation that is already familiar to many people. The new becomes endowed with a history. One disadvantage, however, is that the conditions and potential areas of use for the new phenomenon might differ greatly in reality, which might not be fully understood or appreciated if 'seen' or understood as a traditional camera with traditional functions and norms:

> When the technological media of an artifact changes, some characteristics of the previous media are left behind, others are brought forward intact into the new media, while still others are brought forward in a modified form. In the transition between the non-digital and digital media, a learning process occurs where users employ metaphors from the non-digital representation and process to orient themselves to the novelty of the new media. (Cass and Lauer, 2004, p. 225)

Analogically, I seek to problematise cases where the conceptual lock-in and path dependence becomes *overly* conservative and retrospective, and lead in a direction that frustrates law's relation to society. Where everything new is deemed and evaluated as illegitimate, incomprehensible, or irrelevant; where the 'learning process' is slow; where, for example, work environment regulation is not capable of incorporating new—sometimes transboundary and digital—work environment hazards; or where copyright cannot regulate or stimulate new creative expressions because these are being expressed via other production methods than were available when that specific legal area was largely created. Innovation and entrepreneurship—to take a fairly general example from a study on Skype, Spotify, and TPB (Palmås et al., 2014)—necessarily always end up negotiating their legitimacy in relation to the already established market and actors. One of the mechanisms of path dependence is that at least part of the innovation will be disruptive, since it challenges the order of things as well as already established actors who have not benefited from the challenges. The first disruptors may, however, pave the way for succeeding players, as argued for by Elert et al. (2016) in the case of TPB.

There is a delay in how new phenomena become defined, since they are often interpreted and given meaning based in an already existing historical framework. This delay or conceptual link between the new and the old is sometimes termed, within for instance design research, as *skeuomorph*. A skeuomorph provides us with familiar elements in an unfamiliar domain by seizing upon elements that make new, unknown phenomena appear old and familiar (Gessler, 1998). This recycling or broadened use of a metaphor often occurs quite naturally or, even, out of necessity.

It is at this point that physical 'copies' and copyright law's specific focus on copy reproduction can be analysed, as I do in chapter 5. There, I conclude that for many people, the contemporary concept of copies differs from how it is prescribed in law, and I trace this legal focus on physical copy reproduction to a wholly different sociotechnological context than a digital environment. The limitations that existed then no longer exist in the same manner, as the premises have changed. This has also had an impact on the social norms that have developed around digital media. In other words, what we perceive as right or wrong concerning these artefacts depends on how they are represented, and on how films, music, images, and books are mediated. These conditions contribute to defining the *social norms*, irrespective of what 'sluggish' law prescribes. I conclude in chapter 5 that contemporary copyright law does not regulate 'reality as reality is perceived'. The distance between social conceptualisation and legal conceptualisation is wide enough for the social contract to be broken in this case, at least during the most intensely file-sharing days of the millennium's first decade (cf. Larsson, 2013b, p. 631). I argue that 'culture-bearing products' in actual fact are understood more as a volatile flow than the regulating control-over-copy that copyright is founded on.

THE DIGITAL CHALLENGE TO COPYRIGHT

A common narrative is the one that the development of law is in general conservative and retrospective. Embedded values are long lasting and consequent upon the main principle of predictability (Larsson, 2011c). The problem, however, is how to deal with aspects of law lagging behind social development, when relatively sudden changes in the social structures, perhaps as a result of new technology, emerge and challenge the conservative and often retrospective legal setting (Abel, 1982). The dependence of the path chosen—in law—can sometimes likely be explained by the lock-in effects of the unavoidable use of metaphorical concepts and conceptions. As Roscoe Pound, inspired by Ehrlich

([1913]1936) and founder of the movement of 'sociological jurisprudence', put it a hundred years ago:

> [L]aw has always been dominated by ideas of the past long after they have ceased to be vital in other departments of learning. This is an inherent difficulty in legal science, and it is closely connected with an inherent difficulty in the administration of justice according to law—namely, the inevitable difference in rate of progress between law and public opinion. (Pound, 1910, pp. 25–6)

Cognitive theory teaches us not only that abstract concepts are largely metaphorical but that the metaphor depends on a larger context. This contextuality takes part in a social world that can also be analysed. Meaning is not only built up of the kinds of bodies and social experiences we have, it is framed and constrained by the systematic nature of cognitive processes such as metaphors. This is why the Internet and similar technologies have such vast implications for legal imperatives. Legal imperatives need to be placed in a context of a 'massive cultural tableau', as Winter puts it (2008, p. 375), in order to be comprehensible and understandable. Legislators, too, can only act in terms of the embedded cultural understandings that enable meaning, which Winter describes as 'an important part of any statute is not made by the legislator but is contingent on the pre-existing practices that are conventional for and constitutive of that culture' (Winter, 2008, p. 375).

Copyright law is global, and is entrenched in international treaties as well as in European and Swedish law, which was applicable in the Pirate Bay court case. I argue in a previous chapter that the seemingly insignificant component of copyright that focuses on control of the reproduced copy expresses a notion or conceptual metaphor that encompasses large portions of copyright law—and is currently in conflict with reality. In contrast, and to introduce a duelling combatant to the concept of copies, the concept of *flow* as a metaphorical description of digital distribution can also be discussed. Depending on whether copyright law is worded according to a flow-metaphor or the concept of protection of reproduction, one will arrive at very different results. The Creative Commons (CC) concept, for example, is a way to let creators themselves regulate the use of their creations through licences in relation to what aspects they feel are the most relevant to them—be it just attribution, while allowing also commercial reuse and even remix, or a stricter take allowing only noncommercial reuse, etc. This could be labelled as one way to acknowledge a more flow-like conception of culture, often described through an aim of creating a richer public domain.

When the District court in The Pirate Bay case found that the damages awarded to the plaintiffs were to be at roughly EUR 3,500,000, they largely based their decision in a calculation model based in the number of downloads, and each download represented a set and fixed value, despite the model being dependent on several unreasonable presumptions—for example, equating downloads with paid consumption, or copy-reproduction with theft, despite the latter generally implying a dispossession of some kind, unlike the former—or, at least not by the same standards. By applying traditional analogical concepts and evaluating and categorising the 'digital flow' in terms of physical copies, an essentially different practice arises—one which can generate absurd monetary values if followed to its logical conclusion. The flow conception however by no means indicates that there isn't any viable business model to be applied; on the contrary, it is merely a different kind of business model than the one operating on pay per copy or particular content. I will further develop this in section "'Invent the age! Invent the metaphor!'" below, in terms of an 'openness' metaphor.

There is a lock-in effect in the way metaphors and concepts function beneath the conscious level, as something conservative and difficult to criticise. Bill D. Herman (2008) has convincingly demonstrated that the ownership metaphor dominates the general mental imagery in copyright, and therefore much of the debate—even among theorists that attempt to reconceptualise the problems of digitalisation. Metaphors are convincing conceptual devices that frame phenomena in particular ways, and they can lend a pedagogical and rhetorical advantage to anyone arguing for conceptual links to the ownership of physical property. This, in turn, propagates the concept of physical copies, but also provides an argumentative upper hand by framing the debate in terms of 'theft', 'pirates', 'parasites', 'trespassing,' etc., i.e. actions that are based in an analogue setting containing physical objects, but are metaphorically transferred in order to define new actions linked to digitalisation.

As I have developed elsewhere (Larsson, 2014c), the digitalisation of media entails an ontological change. In other words, the practical conditions in reality within the relevant area have changed. This must be taken into account when understanding the challenges that follow as a result of *practical* copyright changes. The concepts have been endowed with new meaning, or rather; the concepts have colonised new practices and therefore have expanded meaning, as described in chapter 7. For example, Vaidhyanathan explains how access to a work coincides with its use and copying in the digital environment (2001, p. 152). This is what Lessig describes as an exaggerated focus on physical copies (2004, p. 140). Vaidhyanathan explains that this focus on physical copies has led legislators to face difficult choices of either relinquish some control over copying or expand copyright to regulate access and use—despite the chilling effect this might have on creativity, community, and democracy (Vaidhyanathan, 2001, pp. 152–3).

Both Vaidhyanathan and Lessig thereby point out that the content of that which law is intended to regulate has undergone a conceptual change. Copyright law emerges, once again, as an especially salient case study of the inherent conflict between the regulation of phenomena and conditions that have dramatically changed due to technological development. Because the difference is ontological—the films, music, and books are stored and mediated in new ways—regulation also becomes particularly challenged in those areas where it is format-dependent.

A problematic consequence with law being a fixator of processes and organisational settings lies in its relation to being a fixator also of creative expressions with regards to how we understand creativity, authorship, and authenticity. A copyright with long-term protection not only affects the power balances for new innovation—i.e. Spotify and the likes trying to legally obtain a rich catalogue to offer (Palmås et al., 2014)—but also for new forms of creativity, a point not the least argued for by Lessig in *Remix* (2008; cf. Halbert, 2014, pp. 135–8, on 'cultural flows' and the state). Part of this 'fixation' lies in the separation of authenticity and imitation, in the attempts to distinguish between inspiration and plagiarism, for example. This is sometimes renouncing the cultural basis all types of creativity by necessity departs from, or is attached to, and the fact that also creative expressions are technology-dependent, and that with new technological possibilities also the creative expressions change, sometimes challenging the legal conceptions of what should be protected. Halbert addresses this dilemma of authenticity and remix from a more political perspective: 'The quest for the authentic is a search for an elusive idea that is not located in culture. When a property law seeks to put boundaries on a cultural product and export it as a symbol of the nation, it interrupts the otherwise global flow of creativity', and succinctly summarises with the conclusion that 'The fallacy is cultural purity—it is cultural hybridity that is the norm' (2014, p. 138).

If we accept that one purpose of copyright is to stimulate creativity, then the above raises abundantly clear questions concerning the kind of creativity that should be stimulated. Digitalisation has partly shown that an immense amount of cultural expressions exists without any legal protection whatsoever, and partly that a large amount of these expressions are, in fact, counteracted by already existing, and very extensive, protection provided by copyright law. Therefore, copyright law becomes a sort of cultural-political expression that conveys that a certain type of culture (songs, texts, films) is preferable to another (collages, derivative works, mashups, etc.), in which various states, the EU, and other international organisations are involved in both cultural politics and distribution politics (the traditional actors more so than the new ones). Since the notion of incentive is so ingrained, invoking it also leads to a rhetorical advantage for those advocating stronger protection, longer duration of copyright, and simplified legal application. This, in turn, stunts or hinders some cultural expressions, benefits established middlemen, counteracts effectivisation, and risks turning Internet service providers into surveillance officers for the interests of a particular industry.

METAPHORS AND POWER

Metaphor and conceptual frameworks control, to a large degree, how we communicate and therefore how we argue and debate at a fundamental level. There also seems to be a widely spread notion that our thought processes are more objectively anchored than they actually are. Here, a process, or a form of conflict, becomes salient that indicates a power situation. The interested party that successfully establishes the metaphors and concepts that suit their case will be able to lead the discussion in their favour. This is also expressed in how law is formulated, which occurs dialectically with technological development, changed social conditions, etc. In addition, law also adds the fundamentally important aspect of *formalised power* in line with sanctioning metaphors and concepts. Reality must, in fact, be translated to legal language and adapted to its expressions, in order for the legal process to proceed. And in that process, a false sense of security and objectivity may be produced. It could be said that the legal system's legislation, preparatory work, and courts also construct a worldview in themselves or as a sort of mutual understanding of how the world *is*—as is often the case concerning legal norms—at least implicitly of how it *ought* to be. Continuing on the theme of written law, we can conclude that there are preambles that bear witness to a humanist perspective as a sort of overarching concept, and there are an indescribable number of detailed paragraphs that bear witness to concepts concerning more specified factual conditions. Sometimes, however, these small and seemingly insignificant paragraphs inform us of an entire worldview that may fundamentally underlie a globally spread regulation that affects billions of people.

In the transition from an analogical to a digitally based social system, including all the changes this entails for social patterns, work organisation, politics and democracy, nation states and legitimacy, creativity and reproduction—and, not the least, the role of law—the mental imagery, metaphors and their underlying concepts will sometimes be turned upside down, and sometimes remain unperturbed. To better understand the social norms and behaviour currently in change, and the legal and political changes they entail, we need to better understand how language and law—and technology—are related. This also places certain demands on the processes and organisation of science. Some legal components work well in the new context, other less so, and it is important to not become too stuck in old ruts to embrace the new, or to critically analyse and study the old. The objective is to reach a delicate balance where culture is allowed to just happen, thus creating a good interaction between social and legal norms, while simultaneously making room for the developments of new technologies and their social consequences.

As mentioned, we can never be completely free of our tradition-bound conceptualisation of new and abstract phenomena. We need concepts and we need to create understanding of new phenomena. But we must also be aware of this conceptual path dependence, perhaps particularly so in the context of legal development, since law is one of the most powerful institutional tools in society. The case against the founders of TPB shows, through my analysis, that interpreting new phenomena within the context of traditional legislation has a significant enough impact to imprison people and sentence them to life-long indebtedness. Since normativity is linked to how we interpret phenomena, we also have emotions, well established practices, and thoughts on right or wrong already linked to the old concept, the lending domain. Therefore, there is a risk that we shut in the new, stifle it, suffocate it in practices that have already been fixed, within the conventional and the already established. The power aspect of how to (define and) understand the Internet is very relevant also for the proliferation of industries. Different types of industries have stakes in how online media is regulated and how these regulations are enforced. The conflict concerns not only traditional record companies, publishers, and movie industries on the one side and online piracy on the other, but also between the former industries and a newer type that profits from content being posted and shared on their and other, preferably 'open' platforms and websites. Their game is to index it, retain individual data from it in masses, and to make commodities of that data, for example to sell ads (Taylor, 2014). This massive harvesting, analysis and trade of individuals' data is a core business for Google, which also owns YouTube, and the big social media players Facebook and Twitter, but it could in a sense also be argued that this includes the industry that designs and constructs the gadgets we use as well as the services providing us with Internet connectivity. They all benefit from us being constantly connected and preferably stimulating much online traffic, and not necessarily paying per copy or using licenced content. These types of industries therefore benefit from connecting their business models to any metaphor with more positive connotations than what the so-called content industry may offer, such as the need for a 'free and open' Internet.

'INVENT THE AGE! INVENT THE METAPHOR!'

In contemporary, and highly smartphone-and-tablet penetrated society, one could speculate on what metaphor would better capture keyboard-freed Internet connectivity, but it should arguably be seen as an empirical question—how do we understand the Internet today? To what extent is it changing? One could argue that for the touchscreen users we are at the point when the distinction between online and offline has lost its meaning (a recent Swedish study concluded that 85% of the surveyed

18–25-year-olds use the Internet 'virtually all the time' (Larsson et al., 2016)) and perhaps that the keyboard as the main access to the Internet is losing ground. The Internet, as the saying has been for a while now, is also increasingly becoming an Internet 'of things', a development that may require or lead to other metaphorical framings and conceptualisations than before. Multiple meanings could of course exist at the same time—we may envision and understand the Internet differently, even if we may use similar lingual expressions, likely sometimes leading to miscommunication.

Nevertheless, a contemporary, embodied metaphor that has arguably become even more important for the understandings of digital architecture, business models, as well as governmental practices, is that of *open*. It may be explicitly used, such as in *open source, open access, open data*, or *open innovation,* or stand for an inexplicit notion or conception of online environment as an environment that has developed from proprietary business models of selling products copy by copy in streaming services according to a more 'open' flow of media. The latter, building on a 'free' flow, can be seen in subscription models (for example utilised by Spotify and Netflix) or for 'free' services, as in the users being measured and targeted for ads, which shapes the main revenues for global new media players such as Google, Facebook, and Twitter. *Open* is nevertheless too a contemporary metaphor of strong 'discursive resonance', to paraphrase Gillespie's description of platforms (2010).

Metaphorically, it is of interest to see what type of values that may be hidden beneath the 'open' concept. First of all, a reference can be made to Reddy's container schema as an embodied metaphor (see chapter 2; cf. Reddy, 1979) which can apply to abstract concepts such as opposing viewpoints, in that one can have an *open* or a *closed* mind. The discussions on freedom, openness, control, and regulation often emerge when it comes to the Internet and digitalisation. The *open source movement* is a clear example, pushing for the benefits of being able to co-work in primarily software development, as in a movement of 'free and open source' (cf. Söderberg, 2008). *Open access* on the other hand has underscored some of the imbalances of conventional academic publishing, sometimes leading to tax-funded research being locked into the proprietary models of the publishers and sold back to the research community through massive license agreements (Haider and Åström, 2016). The development of open access is challenging this imbalance, albeit not without quite substantial critique with regards to, for example, quality assurance. Similarly, there has in the last few years been what can be called an *open data movement* that has taken on international significance, with government agencies around the world committing to releasing data through government and nongovernmental websites. This initiative is aimed at promoting government accountability and civil participation as well as encouraging commerce and innovation through the release of public sector information (PSI). In the European Union, the push for *open* data has been marked by Directive 2003/98/EC on the reuse of public sector information (the PSI Directive).

A related concept to open data is that of *open innovation*—and both draw from the principal notion that it is not necessarily the initiating party, or data holding party, that holds all the keys to possible innovations for a given project. Open innovation is promoted by, for example, Henry Chesbrough (2003; cf. Munir et al., 2016) in *Open Innovation: The New Imperative for Creating and Profiting from Technology*. The concept of 'open data' comes with a notion of a more passive keeper of the data, where it is assumed that external parties may be innovative based on the available data (passive/active), whereas open innovation signifies a more active approach on both sides, albeit the initiator is structurally 'open' for others' contributions (active/active). The open innovation is sometimes contrasted—quite naturally, concerning this particular embodied concept—by *closed* innovation, often meaning that a company should control the generation of their own production and innovation, and also often in terms of being an 'older' model (Chesbrough, for example, differs between the open and closed 'innovation paradigms' with reference to Thomas Kuhn).

Some quite substantial critique concerning the appraisal for 'openness' in a digital context has been voiced the last few years. For example, Evgeny Morozov in *To Save Everything, Click Here* (2013) points out the concept itself as troublesome—'the ambiguity of a term like "openness" in part explains the confusion, excitement and disappointment generated by various recent campaigns to promote "open government" and liberate "open-government data"' (2013, p. 93)—and discusses the normative content that follows in terms of 'openness and its messiahs'. Similarly, Astra Taylor, in *The People's Platform*, describes *open* as a concept 'capacious enough to contain both the communal and capitalistic impulses central to Web 2.0 while being thankfully free of any socialist connotations', adding that new-media thinkers have 'claimed openness as the appropriate utopian ideal for our time, and the concept has caught on' (Taylor, 2014, p. 21). Further, she argues, 'in tech circles, open systems—like the Internet itself—are always good, while closed systems—like the classic broadcast model—are bad'. Therefore, *open*, being an embodied concept, carries with it normativity for both digital architecture and business models:

> However imprecise the terms are applied, the dichotomy of open versus closed (sometimes presented as freedom versus

> control) provides the conceptual framework that increasingly underpins much of the current thinking about technology, media and culture. (Taylor, 2014, p. 21)

Taylor develops her critique that new media players, now grown into megacorporations, share similar monopolist traits to traditional mass media, but with the rhetorical advantages brought by the 'openness' metaphor. We should by no means be tricked, she argues, into thinking that *open* by necessity means *free*. Instead, this business model is not built around selling specific products but around measuring us, storing and selling our data to the advertising industry. She argues that this 'open model' leads to the distinction between commercial and noncommercial having 'melted away': 'Where there is no distinction between inner and outer, our bonds with family and friends, our private desires and curiosities, all become commodities' (2014, p. 212).

As you know, one of the common features of metaphors, however, is that just as they highlight some aspects of a phenomenon they also hide other aspects of it. Just like 'streamed media' emphasises evanescence and 'fluidity', it also conceals the fact that the files are at some point downloaded to the device, and, just like 'cloud computing', cognitively downplays the physical sides of the necessary data storage. It is not only borderless and easily accessed; it also demands tremendous amounts of energy, cooling systems, and server farms (cf. Hu, 2015). One of the attributes of *open* that makes it such a popular concept at this stage of digitalisation is arguably the same reason we will keep seeing the platform metaphor for many new ventures, often linked to the notion of the 'sharing economy': One of the most signifying aspects in contemporary Internet is a shift from production to distribution, which can be seen in the organisation of some of the latest companies creating a buzz. As Tom Goodwin (3 March 2015) succinctly put it in an editor's note at *Tech crunch*, a technology industry news site:

> Uber, the world's largest taxi company, owns no vehicles. Facebook, the world's most popular media owner, creates no content. Alibaba, the most valuable retailer, has no inventory. And Airbnb, the world's largest accommodation provider, owns no real estate. Something interesting is happening.

Concomitantly, the embodied *open* metaphor seems central both to how businesses develop as well as regulations that apply today and, at least, for the next few years to come.

On liquidity and fear as part of a conceptual process

Much literature on digital challenges to copyright refers to liquid notions of counter-narratives, as pointed out by Fleischer (2010) and described as a 'juridical liquid space' by Dahlberg (2011). This ties on to early conceptions of the Internet, as something to 'surf', for example through Netscape *Navigator*. I, too, describe the counter-conceptions to the legal metaphors in terms of 'flows'. Perhaps the 'liquidity' expresses a counter-conception to LAW IS CONTROL, signifying a state of disruption that may be transitional. Imaginably is this type of somewhat chaotic and uncontrollable conceptualisations significant for early stages in perceivably abstract technologically-dependent development. Because, there are similar metaphorical displays in the conceptual and political struggles in the specific case of big data and human-centric data collection and analysis, debated mostly over the very last few years. These vivid descriptions of data as a fluid, uncontrollable entity possessing great physical power may be a reflection on where in the process of utilising big data our industries and societies currently are, at least in general. This way of understanding massive amounts of human-centric information as related to unpredictability and the difficulty of control and containment may reflect that we still have not gotten quite used to it. In a couple of years' time—when industry, legislations and the services we use are even more mature and unconsciously commonplace—the descriptions and lingual expressions may be related to something more controlled and fixed.

SUM: TECHNOLOGY AND SOCIAL CHANGE

Finally—to categorise according to three approaches that somewhat synthesise the triptych of metaphors, law, and digitalisation that represent the theme of this book—one could for the sake of analysis place digitalisation as the foundation and then project materiality, normativity, and the conceptual onto it. I begin with the material approach, continue with the normative, and conclude with the cognitive, in terms of how to understand abstract phenomena—and consequently how to study these based in a scientific approach, here mainly related to metaphors and cognitive science.

Let us begin by summarising digitalisation's artefacts, including the production of vast amounts of complex data concerning all sorts of human actions and events of our era, continuously piling on in real-time. This can be called *the material aspect of digitalisation*. This aspect is somewhat implicitly touched upon in this book—for example, as the technological element that changes the conditions for, and conceptualisation of, creativity.

Digitalisation's materiality also has a controlling function, a normative aspect in which architecture, design, and algorithms interact and influence human interaction and culture; these are ways in which digitalisation influences and leads us. We can call this *the normative aspect of digitalisation*, and I have emphasised this normativity in terms of an ontological change, exemplified with Vaidhyanathan's account on the eBook collapsing traditional categories of reading a physical book, etc. One could also raise the work of science in the area of understanding law's social role and impact in this context, i.e. the normative aspects of digitalisation, and what implications follow from that we now are quantifiable as well as surveillable to such an extreme degree—for society, law, social norms, and behaviours. Finally, the cognitive or *conceptual aspect of digitalisation* relates to the lexicon of metaphoric terminology that is needed and borrowed through embodiment—the cloud, the cookies, the platforms—but first and foremost the conceptual metaphor dependency of how we understand and conceptualise the abstractions of digitality. A key point in this book being that how a phenomenon is conceptualised has implications for normativity related to the specific conceptualisation. And, quite naturally, how a phenomenon is categorised and understood within law has effects on how it is regulated.

Shifting the focus somewhat from the digital, a sizeable selection of this book is devoted to what could be called *law's conceptual metaphor dependence*. I develop law's dependence on materiality, embodiment, and spatiality mainly in chapter 3 and introduce it at a somewhat more general and theoretical level in chapter 2, but with a constant proximity to law and its links to society and culture—a key issue for sociology of law. I would hope that more sociologists of law pay digitalisation attention from their sociolegal perspective, since it brings in its wake a number of normative and key legal attributes that sociology of law has traditionally studied and theorised around. New developments in methodology could also be discussed—the potential is vast—as well as more theoretical concepts, of course, that might be of analytical use in conjunction with modern cognitive science to understand intellectual property, as in chapter 7. Law is often described as a highly autonomous field and from a scientific perspective, then, it becomes even more important to retain a critical perspective of how law describes the world, if we are to uncover problematic issues in law and its relation to society and technology. Since it is crucial to regulate technology-related social change in order for it to work well, in the balance between the path-dependent retrospective view and the potential of emerging conditions, science, too, must develop its methods, theories, and ways of seeing patterns, forming understandings, and explaining contexts.

Conceptions in the code: a brief epilogue

The Internet is both very young and very old at the same time. Many of the early features and infrastructural characteristics still stand, but still almost everything is different and under constant and innovative change—technologically and conceptually. I still remember the feeling when it became apparent to me that websites could register and follow IP addresses and to some extent reveal who was visiting the site. In terms of Internet history, this was ages ago, when the Internet, to me and my peers, was mostly understood as something of an information source, a sort of chaotic but somewhat magic place to look at through grey, non-flat plastic screens. This was before social media as we know it today, and the Internet was often conceptualised as a sort of anonymous cyberspace, separated from the physical world (of 'flesh and steel'). The

IP-tracing puzzled me, caught me off guard, and challenged my conceptualisation of the Internet and the 'World Wide Web' as something you very much anonymously 'surfed'. Today, billion dollar industries—and many of the services we rely on to be social, to work, to inform ourselves—depend on identification and traceability. The Internet, it seems, has gone from being an access to content, where one uses as a pair of binoculars to from a secret hideout overview a world of highly disparate bits of information, to having become a measuring and profiling tool over the personalised masses, that consequently spend their time answering to this constantly present profiling by reacting to and interacting with the flows provided by peers on a mere handful of more or less monopolistic platforms. From a peephole over a disorganised (cyber)space to a very much ubiquitous and constant connectivity harvested for locations and individual preferences. The conceptualisation of what the Internet is and means must therefore change as well, as a sort of process of negotiation on a sociopolitical, technological, as well as individual level. The shifts create a similar shift in demand on law and regulatory processes. Seen from the perspective drawn together by this book, the regulatory challenges that will define much of our Internet-connected societies to come, arguably, then, lie in at least three non-exclusive categories: first, how do we reconcile or diminish the gap between the one side of social structures that conceptualise culture through metaphors of liquidity and flows and the other side of industries that constantly propagate for the need for property-like 'dominion and control' and, hence, stronger IP rights? And what happens if there is no possible reconciliation? Second, how do you conceptualise and regulate the information asymmetries between individuals and data collecting, analysing and brokering platform parties—not the least relevant in terms of privacy—and, third, how do you conceptualise and regulate the relationship between the traditional content-based industries, where copyright is one of the main elements, and the emerging but now often mature platform-style industries that rely on open flows of content and access to indexing? In this regulation of technology-dependent social change, I argue throughout this book, we must be highly aware that how we conceptualise and understand the new phenomena makes all the difference for the outcome of the regulatory efforts. A thorough analysis of metaphors, that is, can therefore explain some of the legal challenges in digital times.

REFERENCES

Abel, R. L. "Law as Lag: Inertia as a Social Theory of Law." *Michigan Law Review* 80(4) (1982): 785–809.

Adams, M. "Norms, Standards, Rights." *European Journal of Political Economy* 12(2) (1986): 363–375.

Adrian, A. "Copyright Protected Works And The Territoriality Principle: The Pirate Bay Deep-Sixed." *Computer Law & Security Report* 22 (2006): 392–401.

Alter, A. "Your E-book Is Reading You." *The Wall Street Journal*. July 19, 2012. http://online.wsj.com/article/SB10001424052702304870304577490950051438304.html [Last visited March 6 2015].

Altschuller, S., and R. Benbunan-Fich. "Is Music Downloading the New Prohibition? What Students Reveal Through an Ethical Dilemma." *Ethics and Information Technology* 11(1) (2009): 49–56.

Amsterdam, A. G., and J. Bruner. *Minding the Law: How Courts Rely on Storytelling, and How Their Stories Change the Ways We Understand the Law—and Ourselves*. Cambridge, MA: Harvard University Press, 2000.

Andersson, J. "Det Dumma Nätet." In *Efter The Pirate Bay*, edited by Jonas Andersson and Pelle Snickars, 49–72. Stockholm: Kungliga Biblioteket, 2010.

Andersson, J. "The Origins and Impacts of Swedish File-Sharing: A Case Study." *Critical Studies in Peer Production* RS 1(1) (2011): 1–18.

Andersson, T. *Conceptual Polemics. Dialectic Studies of Concept Formation*. Lund: Lund University, 1994.

Andersson Schwarz, J. *Online File-Sharing: Innovations in Media Consumption*. New York and London: Routledge, 2013.

Andersson Schwarz, J., and S. Larsson. "On the Justifications of Piracy: Differences in Conceptualization and Argumentation between Active Uploaders and Other File-Sharers." In *Piracy: Leakages from Modernity*, edited by Arvanitakis James and Fredriksson Martin, 219–239. Los Angeles, CA: Litwin, 2014.

Andersson Schwarz, J., and K. Palmås. "Introducing the Panspectric Challenge: A Reconfiguration of Regulatory Values in a Multiplatform Media Landscape." *Central European Journal of Communication* 6(2) (2013): 219–233.

Andersson, J., and P. Snickars., eds. *Efter The Pirate Bay*. Stockholm: Kungliga Biblioteket, 2010.

Andreasson, J., and K. Schollin. "Goda och onda medhjälpare—Är du socialt adekvat, lille vän? Om Pirate Bay-målets betydelse." *Svensk Juristtidning* 23 (2011): 534–556.

Arewa, Olufunmilayo B. "From J.C. Bach to Hip Hop: Musical Borrowing, Copyright and Cultural Context." *North Carolina Law Review* 84(2) (2006): 547–645.

Asplund, J. *Teorier om framtiden*. Stockholm: Liber, 1979.

Asplund J. *En essä om Gemeinschaft och Gesellschaft*. Göteborg, Sweden: Bokförlaget Korpen, 1991.

Aubert, V. (ed.). *Sociology of Law*. New York: Penguin, 1969.

Aubert, V. *Rättssociologi*. Stockholm: Aldus/Bonnier, 1972.

Aubert, V. *Continuity and Development in Law and Society*. Oslo: Norwegian University Press, 1989.

Auerbach, C. A. "The Relation of Legal Systems to Social Change." *Wisconsin Law Review* (1980): 1227–1340.

Banakar, R. "The Sociology of Law: From Industrialisation to Globalisation." University of Westminster School of Law Research Paper No. 11-03. Lund University, Sweden, 2011.

Banakar R. "Who Needs the Classics?—On the Relevance of Classical Legal Sociology for the Study of Current Social and Legal Problems." In *Retssociologi*, edited by O. Hammerslev and M. R. Madsen, 1–31. Copenhagen: Hans Reitzels Forlag, 2012.

Barlow, J. P. *A Declaration of the Independence of Cyberspace*. San Francisco: Electronic Frontier Foundation, 1996.

Beekhuyzem, J., L. von Hellens, and S. Nielsen. "Underground Online Music Communities: Exploring Rules for Membership." *Online Information Review* 35(5) (2011): 699–715.

Bellia, P. L., P. S. Berman, B. M. Frischmann, and D. G. Post. *Cyberlaw: Problems of Policy and Jurisprudence in the Information Age*. St. Paul, MN: Thomson/West, 2004.

Bentham, J. *The Theory of Legislation*. Edited with an introduction and notes by C.K. Ogden. London: Trübner, 1931 [1864].

Berger, L. L. "What Is the Sound of a Corporation Speaking? How the Cognitive Theory of Metaphor Can Help Lawyers Shape the Law." *Journal of the Association of Legal Writing Directors* 2 (2004): 169–208.

Berger, L. L. "Of Metaphor, Metonymy, and Corporate Money: Rhetorical Choices in Supreme Court Decisions on Campaign Finance Regulation." *Mercer Law Review* 58(3) (2007): 949–990.

Berger, L. L. "How Embedded Knowledge Structures affect Judicial Decision Making: A Rhetorical Analysis of Metaphor, Narrative, And Imagination in Child Custody Disputes." *Southern California Interdisciplinary Law Journal* 18(259) (2009): 259–308.

Berger, L. L. "The Lady, or the Tiger? A Field Guide to Metaphor and Narrative." *Washburn Law Journal* 50(2) (2011): 275–318.

Berne Convention for the Protection for Literary and Artistic Works, last amended in Paris on September 28, 1979.

Bjerre, C. S. "Mental Capacity as Metaphor." *International Journal for the Semiotics of Law* 18 (2005): 101–140.

Bjerre, C. S. "Secured Transactions Inside Out: Negative Pledge Covenants Property and Perfection." *Cornell Law Review* 84 (1999): 305–393.

Black, M. *Models and Metaphors: Studies in Language and Philosophy*. New York: Cornell University Press, 1962.

Black, M. "More About Metaphor." In *Metaphor and Thought*, edited by A. Ortony, 19–43. Cambridge, UK: Cambridge University Press, 1979.

Blavin, J. H., and I. G. Cohen. "Gore, Gibson, and Goldsmith: The Evolution of Internet Metaphors In Law and Commentary." *Harvard Journal of Law and Technology* 16(1) (2002): 265–286.

Bloch-Smith, E. *Judahite Burial Practices and Beliefs about the Dead*. Sheffield, UK: JSOT Press, 1992.

Bogdan, M. "Cyberspace Pirates Walk the Plank: Some Comments on the Swedish Judgment in the Pirate Bay Case." *Masaryk University Journal of Law and Technology* 4 (2010): 113–126.

Bolter, J. D., and R. Grusin. *Remediation: Understanding New Media*. Cambridge, MA: MIT Press, 1999.

Bottomore, T., and P. Goode. *Austro-Marxism*. Texts translated and edited by Bottomore and Goode, with an introduction by Bottomore. Oxford: Clarendon, 1978.

Boyle, J. *Shamans, Software and Spleens: Law and the Construction of the Information Society*. Cambridge, MA: Harvard University Press, 1996.

Boyle, J. "A Politics of Intellectual Property: An Environmentalism for the Net." *Duke Law Journal* 47(1) (1997): 87–116.

Boyle, J. *The Public Domain: Enclosing the Commons of the Mind*. New Haven, CT: Yale University Press, 2008.

Brown, I. "Copyright Technologies and Clashing Rights." In *The SAGE Handbook of Intellectual Property*, edited by M. David and D. Halbert, 567–585. Thousand Oaks: SAGE, 2015.

Bröchner, J., Åström, K. and Larsson, S. "Intention in Hybrid Organizations: The Diffusion of the Business Metaphor in Swedish Laws." *International Journal for the Semiotics of Law* 28(2) (2015): 371–386.

Burkhart, P., and J. Andersson Schwarz. "Piracy and Social Change: Resisting Piracy Cultures." *International Journal of Communication* 9 (2015): 792–797.

Burton, S. J. (ed.). *The Path of the Law and its Influence: The Legacy of Oliver Wendell Holmes, Jr.* Cambridge, UK: Cambridge University Press, 2000.

Cannon, Danny, dir. *Judge Dredd*. 1995. DVD. Los Angeles, CA: Buena Vista Pictures.

Carrier, M. A. "The Pirate Bay, Grokster, and Google." *Journal of Intellectual Property Rights* 15 (2010): 7–18.

Carver, T., and J. Pikalo (eds.). *Political Language And Metaphor, Interpreting and Changing the World*. New York: Routledge, 2008.

Case B 13301-06, April 17, 2009, Stockholm District Court, "The Pirate Bay Case."

Case B 4041-09, November 26, 2010, Svea Court of Appeal, "The Pirate Bay Case."

Cass, K., and T. W Lauer. "Media Transitions. The Cases of Digital Imagery and E-Mail." *Information Technology & People* 17(3) (2004): 252–267.

Castells, M. *The Rise of the Network Society: The Information Age, Economy, Society and Culture*. Vol. 1. Oxford: Blackwell, 1996.

Chesbrough, H. W. *Open Innovation: The New Imperative for Creating and Profiting from Technology*. Boston: Harvard Business School Press, 2003.

Cotterrell, R. *The Sociology of Law: An Introduction*. 2nd ed. London: Butterworths, 1992.

Dahlberg, L. "Pirates, Partisans, and Politico-Juridical Space." *Law and Literature* 23(2) (2011): 262–281.

Damjanović, D, and Bruno De Witte. *Welfare Integration through EU Law: The Overall Picture in the Light of the Lisbon Treaty*. EUI LAW 2008/34. Florence, Italy: European University Institute , 2008.

David, M. and D., Halbert (eds.). *The SAGE Handbook of Intellectual Property*. Thousand Oaks, CA: SAGE, 2015.

DeLanda, M. *Intensive Science and Virtual Philosophy*. New York: Continuum, 2005.

Deleuze, G., and F. Guattari. *A Thousand Plateaus: Capitalism and Schizophrenia*. Minneapolis: University of Minnesota Press, 1987.

De Sousa Santos, B. "Law: A Map of Misreading. Toward a Postmodern Conception of Law." *Journal of Law and Society* 14(3) (1987): 279–302.

De Sousa Santos, B. "Three Metaphors for a New Conception of Law: The Frontier, the Baroque, and the South." *Law & Society Review* 29(4) (1995): 569–584.

Drahos, P., and J. Braithwaite. *Information Feudalism: Who Owns the Knowledge Economy?* London: Earthscan, 2002.

Drulák, P. "Identifying and Assessing Metaphors: Discourse on EU Reform." In *Political Language and Metaphor: Interpreting and Changing the World*, edited by T. Carver and J. Pikalo, 105–118. New York: Routledge, 2008.

Easterbrook, F. H. "Cyberspace and the Law of the Horse." *University of Chicago Legal Forum* 207 (1996): 207–216.

Ehrlich, E. *Fundamental Principles of the Sociology of Law.* Cambridge, MA: Harvard University Press, 1936 (1913).

Eisenstein, E. *The Printing Press as an Agent of Change.* Cambridge, MA: Cambridge University Press, 1980.

Elert, N., M. Henrekson, and J. Lundblad. "Two Sides to the Evasion: The Pirate Bay and the Interdependencies of Evasive Entrepreneurship." *Journal of Entrepreneurship and Public Policy* 5(2) (2016): 176–200.

Engberg, J., and K. Wølch Rasmussen. "Cognition, Meaning Making, and Legal Communication." *International Journal for the Semiotics of Law* 23(4) (2010): 367–371.

Feldman, Y. and J. Nadler. "The Law and Norms of File-Sharing." *The San Diego Law Review* 43(3) (2006): 577–618.

Fredriksson, M. "Skapandets rätt: Ett kulturvetenskapligt perspektiv på den svenska upphovsrättens historia." PhD Thesis in Philosophy, Linköping University. Göteborg, Sweden: Daidalos, 2009.

Fleischer, R. "Femton gastar på död mans kista—om framtidens nätpolitik." In *Efter The Pirate Bay*, edited by Jonas Andersson and Pelle Snickars, 259–280. Mediehistoriskt arkiv. Stockholm: Kungliga biblioteket, 2010.

Fleischer, R. *Musikens politiska ekonomi: Lagstiftningen, ljudmedierna och försvaret av den levande musiken, 1925–2000.* Stockholm: Ink Bokförlag, 2012.

Foucault, M. *Discipline and Punish—The Birth of the Prison.* New York: Penguin, 1991.

Geiger, C. "The Anti-Counterfeiting Trade Agreement and Criminal Enforcement of Intellectual Property: What Consequences for the European Union?" In *Intellectual Property at the Crossroads of Trade*, edited by J. Rosen, 167–181. Northampton, MA: Edward Elgar, 2012.

Gessler, N. *Skeuomorphs and Cultural Algorithms.* Los Angeles: UCLA Anthropology, Computational Evolution and Ecology Group, 1998.

Gibbs R. W, Jr. *Embodiment and Cognitive Science*. Cambridge, UK: Cambridge University Press, 2005.

Gibbs, R. W, Jr. "Evaluating Conceptual Metaphor Theory." *Discourse Processes*, 48(8) (2011): 529–562.

Gibbs, R. W, Jr. "Conceptual Metaphor in Thought and Social Action." In *The Power of Metaphor: Examining its Influence on Social Life*, edited by Michael D. Robinson and Brian P. Meier, 17–40. Washington, DC: American Psychological Association, 2014.

Gillespie, T. *Wired Shut: Copyright and the Shape of Digital Culture*. Cambridge, MA: MIT Press, 2007.

Gillespie, T. "The Politics of 'Platforms.'" *New Media & Society* 12(3) (2010): 347–364.

Gillette, C. P. "Lock-in Effects in Law and Norms." *Boston University Law Review* 78(3) (1998): 813–842.

Glucksberg, S. "How Metaphors Create Categories—Quickly." In *The Cambridge Handbook of Metaphor and Thought*, edited by Raymond W Gibbs Jr. Cambridge, UK: University Press, 2008.

Goodenough, O. R. and G. J. Decker. "Why Do Good People Steal Intellectual Property?" In *Law, Mind and Brain,* edited by M. Freeman M and O. R. Goodenough, 345–372. Aldershot, UK: Ashgate, 2008.

Goodwin, T. "The Battle Is For The Customer Interface." *TechCrunch*. March 3 2015. http://techcrunch.com/2015/03/03/in-the-age-of-disintermediation-the-battle-is-all-for-the-customer-interface/#.lcdm70:u0aR [Last visited August 24 2016].

Gottlieb, H. "Anglicisms and Translation." In *In and Out of English: For Better, For Worse?* edited by G. Anderman and M. Rogers, 161–184. Buffalo, NY: Multilingual Matters, 2005.

Gould, J. B, and S. Barclay. "Mind the Gap: The Place of Gap Studies in Sociolegal Scholarship." *Annual Review of Law and Social Science* 8 (2012): 323–325.

Grace, C. and P. Wilkinson. *Sociological Inquiry and Legal Phenomena*. New York: St. Martin's, 1978.

Gracz, K. "Bridging the Gaps Between Social and Legal Norms Concerning Protection of Intellectual and Artistic Creations: On the Crisis of Copyright Law in the Digital Era." *The Journal of World Intellectual Property* 16(1–2) (2013): 39–57.

Greiffenstern, S. "The Influence of Computers, the Internet and Computer-Mediated Communication on Everyday English." Philosophische Fakultät II, Diss. Berlin. Humboldt-Universität, Logos Verlag, 2010.

Gross, S., J. Bardzell, and S. Bardzell. "Skeu the Evolution: Skeuomorphs, Style, and the Material of Tangible Interactions." *TEI '14: Proceedings of the 8th International Conference on Tangible, Embedded and Embodied Interaction*, 53–60. New York: ACM, 2014.

Haider, J. and F. Åström. "Dimensions of Trust In Scholarly Communication: Problematizing Peer Review in the Aftermath of John Bohannon's 'Sting' in Science." *Journal of the Association for Information Science and Technology* (2016): Early view online. http://onlinelibrary.wiley.com/doi/10.1002/asi.23669/full.

Halbert, D. "Intellectual Property Piracy: The Narrative Construction of Deviance." *International Journal for the Semiotics of Law* 10(28) (1997): 55–78.

Halbert, D. *The State of Copyright: The Complex Relationships of Cultural Creation in a Globalized World*. London: Routledge, 2014.

Halbert, D., and S. Larsson. "By Policy or Design? Privacy in the US in a Post-Snowden World." *Journal of Law, Technology, and Public Policy* 1(2) (2015): 1–17.

Hathaway, O. A. "Path-Dependence in the Law: The Course and Pattern of Legal Change in a Common Law System." *Iowa Law Review* 86(2) (2001): 601.

Hayles, K. N. *How We Became Posthuman: Virtual Bodies in Cybernetics, Literature, and Informatics*. Chicago: The University of Chicago Press, 1999.

Helman, L. "Pull Too Hard and the Rope May Break: On the Secondary Liability of Technology Providers for Copyright Infringement." *Columbia Law School* (2009): 2–56.

Hemmungs Wirtén, E. *No Trespassing: Authorship, Intellectual Property Rights, and the Boundaries of Globalization*. Toronto: University of Toronto Press, 2004.

Hemmungs Wirtén, E. *Cosmopolitan Copyright: Law and Language in the Translation Zone*. Uppsala, Sweden: Meddelanden från Institutionen för ABM vid Uppsala universitet Nr 4, 2011.

Herman, B. D. "Breaking and Entering My Own Computer: The Contest of Copyright Metaphors." *Communication Law and Policy* 13(2) (2008): 231–274.

Hohfeld, W. N. "Some Fundamental Legal Conceptions As Applied in Judicial Reasoning." *Yale Law Journal* 23(1) (1913): 16–59.

Hsu, H. "How the Metaphor of 'the Cloud' Changed Our Attitude Toward the Internet." *The New Yorker*. November 10, 2015. http://www.newyorker.com/books/page-turner/how-the-metaphor-of-the-cloud-changed-our-attitude-toward-the-internet [Last visited 20 March 2016].

Hu, T.-H. *A Prehistory of the Cloud*. Cambridge, MA: MIT Press, 2015.

Hunter, D. "Cyberspace as Place and the Tragedy of the Digital Anticommons." *California Law Review* 91(4) (2003): 439–519.

Hutton, C. "Meaning, Time and the Law: Ex Post and Ex Ante Perspectives." *International Journal for the Semiotics of Law* 22(3) (2009): 279–292.

Hydén, H. *Rättens samhälleliga funktioner.* Lund, Sweden: Studentlitteratur, 1978.

Hydén H., and M. Svensson. "The Concept of Norms in Sociology of Law." In *Law and Society*, edited by P. Wahlgren, 15–32. Scandinavian Studies in Law Series, Stockholm: Institute for Scandinavian Law (53), 2008.

INFOSOC Directive: Directive 2001/29/EC of the European Parliament and of the Council of 22 May 2001 on the Harmonisation of Certain Aspects of Copyright and Related Rights in the Information Society. EUR-LEX. http://eur-lex.europa.eu/legal-content/en/TXT/?uri=CELEX:32001L0029.

Jakobsson, P. "Obegränsad kommunikation, obegränsad konkurrens." In *Efter The Pirate Bay*, edited by Jonas Andersson and Pelle Snickars, 87–104. Mediehistoriskt arkiv. Stockholm: Kungliga biblioteket, 2010.

Jakobsson P. *Öppenhetsindustrin.* Doctoral Dissertationss 65. Örebro Studies in Media and Communication 13, Södertörn Doctoral Dissertations, 2012.

Jensen, C. "The More Things Change, the More They Stay the Same: Copyright, Digital Technology, and Social Norms." *Stanford Law Review* 56 (2003): 531–570.

Johansson, D. "När alla blir radiokanaler och frisörer." In *Efter The Pirate Bay*, edited by Jonas Andersson and Pelle Snickars, 105–123. Mediehistoriskt arkiv. Stockholm: Kungliga biblioteket, 2010.

Johnson, D. R., and D. Post. "Law and Borders—The Rise of Law in Cyberspace." *Stanford Law Review* 48 (1996): 1367.

Johnson, M. L. *The Body of the Mind: The Bodily Basis of Meaning, Imagination, and Reason.* Chicago: University of Chicago Press, 1987.

Johnson, M. L. *Moral Imagination: Implications of Cognitive Science for Ethics.* Chicago: University of Chicago Press, 1993.

Johnson M. L. "Law Incarnate." *Brooklyn Law Review* 67(4) (2002): 949–962.

Johnson, M. L. "Mind, Metaphor, Law." *Mercer Law Review* 58 (2007): 845–868.

Joo, T. W. "Contract, Property and the Role of Metaphor in Corporations Law." *UC Davis Daniel J. Dykstra Memorial Corporate Governance Symposium: UC Davis Law Review* 35(779) (2001): 1–59.

Kahn-Freund, O. "Introduction." In *The Institutions of Private Law and their Social Functions* by K. Renner. New Brunswick, NJ: Transaction, 2010 (1949).

Kann, R. A. "Karl Renner (December 14, 1870–December 31, 1950)." *The Journal of Modern History* 23(3) (1951): 243–249.

Karaganis, J., V. Grassmuck, and L. Renkema. *Copy Culture in the US and Germany*. New York: Columbia University, The American Assembly, 2012.

Kenney, M., and J. Zysman. "The Rise of the Platform Economy." *Issues in Science and Technology* Spring issue (2016): 61–69.

Kingery, W. D. *The Social and Cultural Contexts of New Ceramic Technologies*. Westerville, OH: American Ceramic Society, 1993.

Kinsey, R. "Karl Renner on Socialist Legality." In *Legality, Ideology, and the State*, edited by D. Sugarman, 11–42. London: Academic.

Kirchberger, C. *Cyber Law in Sweden*. Dordrecht, The Netherlands: Kluwer Law International, 2011.

Klang, M. "A Critical Look at the Regulation of Computer Viruses." *International Journal of Law and Information Technology* 11(2) (2003): 162–183.

Knowles, M., and R. Moon. *Introducing Metaphor*. London: Routledge, 2006.

Knutagård, M. *Skälens fångar: Hemlöshetsarbetets organisering, kategoriseringar och förklaringar*. Lund University. Lund, Sweden: Egalité, 2009.

Kövecses, Z. *Metaphor in Culture: Universality and Variation*. Cambridge, UK: Cambridge University Press, 2005.

Kövecses, Z. "Conceptual Metaphor Theory: Some Criticisms and Alternative Proposals." *Annual Review of Cognitive Linguistics* 6 (1) (2008): 168–184.

Kravets, David. "ISPs to Disrupt Internet Access of Copyright Scofflaws." *Wired*, July 7, 2011.

Kullenberg, C. *Det Nätpolitiska Manifestet*. Stockholm: Ink Bokförlag, 2010.

Lakoff, G. "A Figure of Thought." *Metaphor and Symbolic Activity* 1(3) (1986): 215–225.

Lakoff, G. *Women, Fire and Dangerous Things: What Categories Reveal about Mind*. Chicago, IL: University of Chicago Press, 1987.

Lakoff, G. "The Contemporary Theory of Metaphor." In *Metaphor and Thought*, edited by A. Ortony. 2nd edition. Cambridge, UK: Cambridge University Press, 1993.

Lakoff, G. *Don't Think of an Elephant: Know Your Values and Reframe The Debate*. White River Junction, VT: Chelsea Green, 2005.

Lakoff, G., and M. Johnson. *Metaphors We Live By*. Chicago: University of Chicago Press, 1980.

Lakoff, G., and M. Johnson. *Philosophy in the Flesh: The Embodied Mind and Its Challenge to Western Thought*. New York: Basic Books, 1999.

Larsson, S. "459 miljarder kronor—om metaforer, flöden & exemplar." In *Efter The Pirate Bay*, edited by J. Andersson and P. Snickars, 173–195. Mediehistoriskt arkiv. Stockholm: Kungliga biblioteket, 2010.

Larsson, S. "Den stigberoende upphovsrätten. Om konsekvenserna av rättslig inlåsning i en digital tid." *Retfærd, Nordic Journal of Law and Justice* 4(135) (2011a): 122–146.

Larsson, S. "Metaphors and Norms: Understanding Copyright Law in a Digital Society." PhD Thesis, Lund Studies in Sociology of Law, Lund University, Sweden, 2011b.

Larsson, S. "The Path Dependence of European Copyright." *SCRIPT:ed. A Journal of Law, Technology, and Society* 8(1) (2011c): 8–31.

Larsson, S. "Conceptions in the Code: What 'The Copyright Wars' Tells about Creativity, Social Change and Normative Conflicts in the Digital Society." *Societal Studies*, 4(3) (2012a): 1009–1030.

Larsson S. "Metaforerna och Rätten." *Retfærd Nordic Journal of Law and Justice* 2 (137) (2012b): 69–93.

Larsson, S. "'No man Is an Island': Why the 'Solitary Genius' is a Too Narrow Approach on Creativity in a Digital Context." *Linguaculture: International Journal of the Iaşi Linguaculture Centre for (Inter)cultural and (Inter)lingual Research* 2 (2012c): 615–634.

Larsson, S. "Conceptions, Categories, and Embodiment—Why Metaphors Are of Fundamental Importance for Understanding Norms." In *Social and Legal Norms: Towards a Socio-Legal Understanding of Normativity*, edited by M. Baier, 121–141. Farnham, UK: Ashgate. 2013a.

Larsson, S. "Copy Me Happy—The Metaphoric Expansion of Copyright in a Digital Society." *International Journal for the Semiotics of Law* 26(3) (2013b): 615–634.

Larsson, S. "Metaphors, Law and Digital Phenomena: The Swedish Pirate Bay Court Case." *International Journal of Law and Information Technology* 21(4) (2013c): 354–379.

Larsson, S. "Rätten som Rum, Kropp och Kontext: Det Konceptuella Förtingligandets Rättsvetenskapliga Relevans." *Tidsskrift for Rettsvitenskap* 126(3) (2013d): 343–363.

Larsson, S. "Sociology of Law in a Digital Society—A Tweet from Global Bukowina." *Societas/Communitas* 15(1) (2013e): 281–295.

Larsson, S. "Conceptions of Copyright in a Digital Context. A Comparison between French and American File-Sharers." *Lexis—E-journal in English Lexicology* (8) (2014a): 89–102.

Larsson, S. "Justice 'Under' Law—The Bodily Incarnation of Legal Conceptions Over Time." *International Journal for the Semiotics of Law* 27(4) (2014b): 613–626.

Larsson, S. "Karl Renner and (Intellectual) Property—How Cognitive Theory Can Enrich a Sociolegal Analysis of Contemporary Copyright." *Law & Society Review* 48(1) (2014c): 3–33.

Larsson, S. *Metaforerna, rätten & det digitala: Hur gamla begrepp styr förståelsen av nya fenomen*. Malmö, Sweden. Gleerups Utbildning AB, 2015.

Larsson, S. and H. Hydén. "Law, Deviation and Paradigmatic Change: Copyright and Its Metaphors." In *Technology for Facilitating Humanity and Combating Social Deviations: Interdisciplinary Perspectives*, edited by M. Vargas Martin, et al., 188–208. Hershey, PA: Information Science Research, 2010.

Larsson, S. and M. Svensson. "Compliance or Obscurity? Online Anonymity as a Consequence of Fighting Unauthorised File-Sharing." *Policy & Internet* 2(4) (2010): 77–105.

Larsson, S., L. Svensson, and H. Carlsson. "Digital Consumption and Over-Indebtedness Among Young Adults in Sweden." LUii Reports, Lund & Landskrona. Sweden: Lund University. http://lup.lub.lu.se/record/8838699. Retrieved August 16, 2016.

Larsson, S., M. Svensson, and M. de Kaminski. "Online Piracy, Anonymity and Social Change—Innovation through Deviance." *Convergence: The International Journal of Research into New Media Technologies* 19(1) (2012a): 95–114.

Larsson, S., M. Svensson, M. de Kaminski, K. Rönkkö, K, and J. Alkan Olsson. "Law, Norms, Piracy and Online Anonymity—Practices of De-Identification in the Global File Sharing Community." *Journal of Research in Interactive Marketing* 6(4) (2012b): 260–280.

Larsson, S., M. Svensson, P. Mezei, and M. de Kaminski. "The Digital IP Challenge Revisited—File-Sharing and Copyright Development in Hungary." *WIPO Journal* 5(2) (2014a): 176–188.

Larsson, S., S. Wnukowska-Mtonga, M. Svensson, and M. de Kaminski. "Parallel Norms: File-sharing and Contemporary Copyright Development in Australia." *Journal of World Intellectual Property* 17(1–2) (2014b): 1–15.

Laski, H. J. "Review of Karl Renner's *The Institutions of Private Law*." *The Modern Law Review* 13 (1950): 389–392.

Leader, S. "Participation and Property Rights." *Journal of Business Ethics* 21(2–3) (1999): 97–109.

Ledendal, J., and S., Larsson. "Ett rättsligt perspektiv på övervakningstrenden: Datalagringsdirektivets underkännande." In *DigiTrust: Tillit i det Digitala. Tvärvetenskapliga perspektiv från ett forskningsprojekt*, edited by S. Larsson and P. Runeson, 69–76. Lundv, Sweden: The Pufendorf Institute for Advanced Studies, Lund University, 2014.

Lemley, M. A. "Property, Intellectual Property, and Free Riding." *Texas Law Review* 83 (2005): 1031.

Lessig, L. *Code and Other Laws of Cyberspace*. New York: Basic, 1999a.

Lessig, L. "The Law of the Horse: What Cyberlaw Might Teach." *Harvard Law Review* 113 (1999b): 501–546.

Lessig, L. *The Future of Ideas: The Fate of the Commons in a Connected World.* New York: Random House, 2002.

Lessig L. *Free Culture: How Big Media Uses Technology and the Law to Lock Down Culture and Control Creativity.* New York: Penguin, 2004.

Lessig, L. *Code Version 2.0.* New York: Basic, 2006.

Lessig L. *Remix: Making Art and Commerce Thrive in the Hybrid Economy.* New York: Penguin, 2008.

Lewen, U. M. Internet File-Sharing: Swedish Pirates Challenge the U.S. *Cardozo Journal of International and Competitive Law* 16 (2008): 173–205.

Lindgren, S. "Introducing Connected Concept Analysis: A Network Approach to Big Text Datasets." *Text & Talk: An Interdisciplinary Journal of Language, Discourse & Communication Studies* 36(3) (2016): 341–362.

Litman, J. *Digital Copyright*. New York: Prometheus, 2001.

Locke, J. *An Essay Concerning Human Understanding.* Edited by Peter Nidditch. Oxford: Oxford University Press 1975 (1690).

Loughlan, P. "Pirates, Parasites, Reapers, Sowers, Fruits, Foxes ... The Metaphors of Intellectual Property." *Sydney Law Review* 28(2) (2006): 211–226.

Loughlan, P. "'You Wouldn't Steal a Car': Intellectual Property and the Language of Theft." *European Intellectual Property Review* 29 (2007): 401–405.

Luhmann, N. *Rechtssoziologie.* Reinbek bei Hamburg, Germany: Rowohlt, 1972.

Lundblad, N. "Law in the Noise Society: ICT-Policy Making And Societal Models." PhD Thesis in Informatics, IT University, Gothenburg, Sweden, 2007.

MacDonald, R. A. "'Social and Economic Control through Law'—Review of Karl Renner's *The Institutions of Private Law and Their Social Functions.*" *Chitty's Law Journal* (25) (1977): 7–18.

MacLeish, A. "Hypocrite Auteur" from *Collected Poems 1917–1982.* Boston: Houghton Mifflin, 1985

Mahoney, J. "Path-Dependence in Historical Sociology." *Theory and Society* 29(4) (2000): 507–548.

Mathiesen, T. *Rätten i samhället: En introduktion till rättssociologin*, [Original title: *Retten i samfunnet*]. Lund: Studentlitteratur, 2005.

Mattsson, J. and S. Larsson. "Juridikens dolda tankestrukturer: Observationer av metaforbruk i den svenska civilrätten." *Tidsskrift for Rettsvitenskap* 127(5) 2014: 512–534.

Mayer-Schönberger, V., and K. Cukier. *Big Data—A Revolution That Will Transform How We Live, Work, and Think.* Boston: Houghton Mifflin Harcourt, 2013.

McCoy, M. D., and J. B. Needham, II. "Cybertheft: Will Copyright Law Prevent Digital Tyranny on the Superhighway?" *Wake Forest Law Review* 30 (1995): 169.

McLeod, K. *Freedom of Expression. Resistance and Repression in the Age of Intellectual Property.* Minneapolis: University of Minnesota Press, 2007.

McManus, J. J. "The Emergence and Non-Emergence of Legislation." *British Journal of Law and Society* 5(2) (1978): 185–201.

Mellinkoff, D. *The Language of the Law.* Boston: Little, Brown & Co., 1963.

Mendis, D. "Digital Economy Act 2010: Fighting a Losing Battle? Why the 'Three Strikes' Law Is Not the Answer to Copyright Law's Latest Challenge." *International Review of Law, Computers and Technology* 27 (1–2) (2013): 60–84.

Miegel, F., and T. Olsson, "From Pirates to Politicians: The Story of the Swedish File Sharers Who Became a Political Party." In *Democracy, Journalism, and Technology: New Developments in an Enlarged Europe,* edited by Nico Carpentier et al. Researching and Teaching Communication Series 4. Tartu, Estonia: Tartu University Press, 2008.

Moohr, G. The Crime of Copyright Infringement: An Inquiry Based on Morality, Harm, and Criminal Theory. *Boston University Law Review* 83 (2003): 731–783.

Morozov, E. *To Save Everything, Click Here: The Folly of Technological Solutionism.* New York: Public Affairs, 2013.

Morra, L. "New Models for Language Understanding and the Cognitive Approach to Legal Metaphors." *International Journal for the Semiotics of Law* 23(4) (2010): 387–405.

Moser, K. "Metaphor Analysis in Psychology—Method, Theory, and Fields of Application." *FQS Forum: Qualitative Social Research* 1(2) (2001). http://www.qualitative-research.net/index.php/fqs/article/view/1090/2388. Accessed July 26, 2016.

Munir, H., K. Wnuk, and P. Runeson. "Open Innovation in Software Engineering: A Systematic Mapping Study." *Empirical Software Engineering* 21(2) (2016): 684–723.

Murrow, G. B., and R. W. Murrow. "A Biosemiotic Body of Law: The Neurobiology of Justice." *International Journal for the Semiotics of Law* 26(2) (2012): 275–314.

Nelken, D. 2009. "The 'Gap Problem' in Sociology of Law: A Theoretical Review." In *Beyond Law in Context: Developing a Sociological Understanding of Law,* edited by D. Nelken. Farnham, UK: Ashgate, 1981.

Nelken, D. "The Gap Problem in Sociology of Law." *Windsor Yearbook of Access to Justice* (1981): 35–62.

Nerlich, B. and D. D. Clarke. "Mind, Meaning and Metaphor: The Philosophy and Psychology of Metaphor in 19th-Century Germany." *History of the Human Sciences* 14(2) (2001): 39–61.

Obasogie, O. K. "Do Blind People See Race? Social, Legal, and Theoretical Considerations." *Law & Society Review* 44(3–4) (2010): 585–616.

Palfrey, J., and U. Gasser. *Born Digital: Understanding the First Generation of Digital Natives*. New York: Basic, 2008.

Palmås, K. *En liten bok om slem*. Stockholm: Federa, 2009.

Palmås, K. "The Pirate Bay-bacillen—tre spekulationer." In *Efter The Pirate Bay,* edited by Jonas Andersson and P. Snickars, 196–211. Mediehistoriskt arkiv. Stockholm: Kungliga biblioteket, 2010.

Palmås, K., J. Andersson Schwarz, and S. Larsson. "The Liability of Politicalness: Legitimacy and Legality in Piracy-Proximate Entrepreneurship." *International Journal of Entrepreneurship and Small Business* 22(4) 2014: 408–425.

Patry, W. *Moral Panics and the Copyright Wars*. Oxford: Oxford University Press, 2009.

Patry, W. *How to Fix Copyright*. Oxford: Oxford University Press, 2012.

Peccarelli, A. M. "The Meaning of Justice." *The Journal of the DuPage County Bar Association* 12 (March 2000). Web archive: http://www.dcba.org/mpage/vol120300art1.

Peczenik, A. *Vad är rätt? Om demokrati, rättssäkerhet, etik och juridisk argumentation*. Stockholm: Fritzes, 1995.

Peters, G. B. *Institutional Theory in Political Science*. 2nd edition. London: Continuum, 2005.

Pierson, P. "*The American Political Science Review* 94(2) 2000: 251–268.

Pierson, P., and T. Skocpol. "Historical Institutionalism in Contemporary Political Science." In *Political Science: The State of the Discipline,* edited by I. Katznelson and H. V. Milner, 693–721. New York: Norton, 2002.

Posner, R. A. "Past-Dependency, Pragmatism, and a Critique of History in Adjudication and Legal Scholarship." *University of Chicago Law Review* 67(3) (2000): 573–606.

Pound, R. "Law in Books and Law in Action." *American Law Review* 44 (1910): 12–36.

Puschmann, C., and J. Burgess. "Big Data, Big Questions. Metaphors of Big Data." *International Journal of Communication* 8 (2014): 1690–1709.

Reddy, M. "The Conduit Metaphor: A Case of Frame Conflict in our Language about Language." In *Metaphor and Thought,* edited by A. Ortony, 284–324. Cambridge, MA: Cambridge University Press. 1979.

Reed, C. *Making Laws for Cyberspace*. Oxford: Oxford University Press. 2012.

Reid, R. March 20, 2012. Blog post. "The numbers behind the copyright math." http://blog.ted.com/2012/03/20/the-numbers-behind-the-copyright- math/.

Reid, R. The $8 Billion iPod. TED. March 21, 2012. https://www.ted.com/speakers/rob_reid [Last visited August 27, 2012].

Renner, K. *Marxismus: Krieg und Internationale*. 2nd ed. Stuttgart, Germany: J. Dietz, 1918.

Renner, K. *The Institutions of Private Law and Their Social Functions*. New Brunswick, NJ: Transaction Publishers, 2010 [1949].

Ritchie, D. L. *Metaphor*. Cambridge, UK: Cambridge University Press. 2013.

Ritchie D. T. "Who Is on the Outside Looking In and What Do They See? Metaphors of Exclusion in Legal Education," *Mercer Law Review* 58(3) (2007): 991–1019.

Robson, P. "Renner Revisited." In *Perspectives in Jurisprudence*, edited by Elspeth Atwooll, 221–235. Glasgow, Scotland: Glasgow University Press, 1977.

Rome Convention: The International Convention for the Protection of Performers, Producers of Phonograms and Broadcasting Organizations. October 26, 1961.

Rosch, E. "Cognitive Representation of Semantic Categories." *Journal of Experimental Psychology* 104(3) (1975): 192–233.

Rosch, E. "Principles of Categorisation." In *Cognition and Categorization*, edited by E. Rosch and B. Lloyds, 27–48. Hillsdale, NJ: Lawrence Erlbaum, 1978.

Rose, M. *Authors and Owners: The Invention of Copyright*. Cambridge, MA: Harvard University Press, 1993.

Russ, William A, Jr. "Radical Disfranchisement in Texas, 1867–70." *The Southwestern Historical Quarterly* 38(1) (1934): 40–52.

Samuelson, P. "Implications of the Agreement on Trade Related Aspects of Intellectual Property Rights for Cultural Dimension of National Copyright Laws." *Journal of Cultural Economics* 23(1) (1999): 95–107.

Samuelson, P., and T. Wheatland. "Statutory Damages in Copyright Law: A Remedy in Need of Reform." *William & Mary Law Review* 51 (2009): 439–511.

Schön, D. A. "Generative Metaphor." In *Metaphor and Thought*, edited by A. Ortony. Cambridge, UK: Cambridge University Press, 1979.

Schottenius Cullhed, S. *Proba the Prophet: Studies in the Christian Virgilian Cento of Faltonia Betitia Proba*. Göteborg, Sweden: Gothenburg University, 2012.

Schultz, M. F. "Copynorms: Copyright and Social Norms." In: *Intellectual Property and Information Wealth: Issues and Practices in the Digital Age*. Edited by Peter K. Yu. Greenwood: Praeger, 2007.

Shannon, P. "Book Reviews": *The Institutions of Private Law and Their Social Functions* by Karl Renner (ed. O. Kahn-Freund, trans. A. Schwarzschild). *Journal of Sociology* 13(3) (1977): 246–264.

Strahilevitz, L. J. "Charismatic Code, Social Norms, and the Emergence of Cooperation on the File-Swapping Networks." *Virginia Law Review* 89(3) (2003a): 505–595.

Strahilevitz, L. J. "Social Norms from Close-Knit Groups to Loose-Knit Groups." *The University of Chicago Law Review* 70(1) (2003b): 359–372.

Svensson, M. *Sociala normer och regelefterlevnad: Trafiksäkerhetsfrågor ur ett rättssociologiskt perspektiv* (Social Norms and the Observance of Law). Doctoral Dissertation, Lund University, Lund Studies in Sociology of Law 28, 2008.

Svensson, M. "Norms in Law and Society." *Social and Legal Norms: Towards a Socio-legal Understanding of Normativity*. Farnham, UK: Ashgate. 2013.

Svensson, M., and S. Larsson. "Social Norms and Intellectual Property: Online Norms and the European Legal Development." Lund, Sweden: Lund University, Research Report in Sociology of Law, 2009.

Svensson, M, and S. Larsson. "Intellectual Property Law Compliance in Europe: Illegal File-Sharing and the Role of Social Norms." *New Media & Society* 14(7) (2012): 1147–1163.

Svensson, M., S. Larsson, and M. de Kaminski. "The Research Bay—Studying the Global File-Sharing Community." In *Intellectual Property in Context: Towards a Law and Society Perspective*, edited by William Gallagher and Halbert Debora. Cambridge, MA: Cambridge University Press. Forthcoming.

Svensson, M., S. Larsson, and M. de Kaminski. "Professionalization, Gender and Anonymity in the Global File Sharing Community." In *Piracy Effect*, edited by R. Braga and G. Caruso, 1–8. Italy: Cinergie.it, 2013.

Sweetser, E. *From Etymology to Pragmatics*. Cambridge, UK: Cambridge University Press, 1990.

Söderberg, J. *Hacking Capitalism—The Free and Open Source Software Movement*. London: Routledge, 2008.

Taylor, A. *The People's Platform: Taking Back Power and Culture in the Digital Age*. New York: Metropolitan Books/Henry Holt, 2014.

Tehranian, J. "Infringement Nation: Copyright Reform and the Law/Norm Gap." *Utah Law Review* 3 (2007): 537–551.

Tiersma, P. M., and L. M. Solan. "Cops and Robbers: Selective Literalism in American Criminal Law." *Law & Society Review* 38 (2004) 229–266.

Touloumis, Tara. "Buccaneers and Bucks from the Internet: Pirate Bay and the Entertainment Industry." *Seton Hall Journal of Sports and Entertainment Law* 19 (2009): 253–281.

Treviño, J. "Introduction to the Transaction Edition." In *The Institutions of Private Law and their Social Functions* by Karl Renner. New Brunswick, NJ: Transaction Publishers, 2010.

Treviño, A. J. *The Sociology of Law. Classical and Contemporary Perspectives*. New Brunswick, NJ: Transaction Publishers, 2008.

Trim, R. *Metaphor Networks: The Comparative Evolution of Figurative Language*. New York: Palgrave Macmillan, 2007.

TRIPS: Agreement on Trade-Related Aspects of Intellectual Property Rights signed in Marrakech, Morocco on April 15 1994. https://www.wto.org/english/tratop_e/trips_e/t_agm0_e.htm. Retrieved July 27, 2016.

Tsai, R. L. "Fire, Metaphor, and Constitutional Myth-Making." *Georgetown Law Journal* 93(1) 2004: 181–239.

Vaidhyanathan, S. *Copyrights and Copywrongs: The Rise of Intellectual Property and How It Threatens Creativity*. New York: New York University Press, 2001.

Vaidhyanathan, S. "The Rise of the Cryptopticon." *The Hedgehog Review* 17(1) (2015): 71–85.

van den Boomen, M. *Transcoding the Digital: How Metaphors Matter in New Media*. Amsterdam: Institute of Network Cultures, 2014.

Volgsten, U. *Musiken, medierna och lagarna: Musikverkets historia och etablerande av en idealistisk upphovsrätt*. Möklinta, Sweden: Gidlunds förlag, 2012.

von Lewinski, S. *International Copyright Law and Policy*. Oxford: Oxford University Press, 2008.

Walter, J., and J. Helmig. "Discursive Metaphor Analysis: (De) Construction(S) of Europe and Technology Studies." In *Political Language and Metaphor: Interpreting and Changing the World*, edited by T. Carver and J. Pikalo. 119–131. New York: Routledge, 2008.

Whitehouse, L. "Making the Case for Sociolegal Research in Land Law: Renner and the Law of Mortgage." *Journal of Law and Society* 37 (2010): 545–568.

Wilken, R. An Exploratory Comparative Analysis of the Use of Metaphors in Writing on the Internet and Mobile Phones." *Social Semiotics* 23(5) (2013): 632–647.

Wingrove, T., A. L. Korpas, and V. Weisz. "Why Were Millions Of People *Not* Obeying the Law? Motivational Influences on Non-Compliance with the Law in the Case of Music Piracy." *Psychology, Crime & Law* 17 (3) (2010): 261–276.

Winter, S. L. *A Clearing in the Forest: Law, Life, and Mind*. Chicago: University of Chicago Press, 2001.

Winter, S. L. "Re-Embodying Law." *Mercer Law Review* 58 (2007): 870–897.

Winter, S. L. "What is the 'Color' of Law?" In *The Cambridge Handbook of Metaphor and Thought*, edited by R. W. Gibbs Jr. Cambridge, UK: Cambridge University Press, 2008.

Wong, M. W. S. "Cyber-Trespass and 'Unauthorized Access' as Legal Mechanisms of Access Control: Lessons from the US Experience." *International Journal of Law and Information Technology* 15(1) (2006): 90–128.

Woodmansee, M. "The Genius and the Copyright: Economic and Legal Conditions of the Emergence of the 'Author.'" In *Special Issue: The Printed Word and the Eighteenth Century*. Edited by Raymond Birn. *Eighteenth-Century Studies* 17(4) (1984): 425–448.

Woodmansee, M. *The Author, Art, and the Market: Rereading the History of Aesthetics*. New York: Columbia University Press, 1994.

Woodmansee, M. "On the Author Effect: Recovering Collectivity." In *The Construction of Authorship: Textual Appropriation in Law and Literature*, edited by M. Woodmansee and P. Jaszi, 15–29. Durham, NC: Duke University Press, 1994.

Wormbs, N. "Det digitala imperativet." In *Efter The Pirate Bay*, edited by J. Andersson and P. Snickars, 140–150. Mediehistoriskt arkiv. Stockholm: Kungliga biblioteket, 2010.

Yar, M. "The Rhetorics and Myths of Anti-Piracy Campaigns: Criminalization, Moral Pedagogy and Capitalist Property Relations in the Classroom." *New Media & Society* 10(4) (2008): 605–623.

Yu, N. "Metaphor from Body and Culture." In *The Cambridge Handbook of Metaphor and Thought*, edited by R. W. Gibbs, Jr., 247–262. New York: Cambridge University Press, 2008.

Yu, P. K. "Digital Copyright and Confuzzling Rhetoric." *Vanderbilt Journal of Entertainment and Technology Law* 13 (2011): 881–939.

Zittrain, J. "The Generative Internet." *Harvard Law Review* 119 (2006): 1974–2040.

Zittrain, J. *The Future of the Internet, and How to Stop It*. London: Penguin, 2008.

INDEX